Fuel Cells, Engines and Hydrogen

Fuel Cells, Engines and Hydrogen

An Exergy Approach

Frederick J. Barclay
C.Eng., C.Phys.,
F.I.Mech.E., F.I.E.E., F.Inst.P.

John Wiley & Sons, Ltd

Other Wiley Editorial Offices

John Wiley & Sons Inc., 111 River Street, Hoboken, NJ 07030, USA

Jossey-Bass, 989 Market Street, San Francisco, CA 94103-1741, USA

Wiley-VCH Verlag GmbH, Boschstr. 12, D-69469 Weinheim, Germany

John Wiley & Sons Australia Ltd, 42 McDougall Street, Milton, Queensland 4064, Australia

John Wiley & Sons (Asia) Pte Ltd, 2 Clementi Loop #02-01, Jin Xing Distripark,
Singapore 129809

John Wiley & Sons Canada Ltd, 22 Worcester Road, Etobicoke, Ontario, Canada M9W 1L1

Wiley also publishes its books in a variety of electronic formats. Some content that appears
in print may not be available in electronic books.

Library of Congress Cataloging in Publication Data

Barclay, Frederick J.
    Fuel cells, engines, and hydrogen : an exergy approach / Frederick J. Barclay.
      p. cm.
    Includes bibliographical references and index.
    ISBN 0-470-01904-2 (cloth : alk. paper)
    1. Fuel cells. 2. Hydrogen as fuel. I. Title.

    TK2931.B38 2006
    621.31′2429—dc22   2006000914

British Library Cataloguing in Publication Data

A catalogue record for this book is available from the British Library

ISBN-13 978-0-470-01904-7 (HB)
ISBN-10 0-470-01904-2 (HB)

Typeset in 10.5/13pt Sabon by Integra Software Services Pvt. Ltd, Pondicherry, India
Printed and bound in Great Britain by TJ International, Padstow, Cornwall
This book is printed on acid-free paper responsibly manufactured from sustainable forestry
in which at least two trees are planted for each one used for paper production.

This book is dedicated to my wife, without whom it could not have been written. The dedication poem below is written in a mixture of Buchan Claik (Buchan dialect) and Lallans (Lowland Scots), as befits a marriage between a husband from Edinburgh, but of Glaswegian upbringing, and a wife from Aberdeen. In Buchan Claik loons (m) and quines (f) are marriageable young people.

My Buchsburn Quine

O Mary, it's lang syne we click'd,
While touching over teacups,
I tak't ye fae a jilted loon,
Sair sick doon in his stommick.

My Buchsburn quine had black dark hair,
That spark'd a'twixt the bed sheets,
Glintin' the way tae lovers rites,
That spritely gint the hert leap.

The potter's clay from high she thumps,
Upon the caring plaster,
Wi' micht an main, tae get it richt,
Tae form, and fire, and gloster.

My Buchan quine, ma ain guidwife,
In the game o'life, a winner,
O'er wrongs and rights, and deil made plights,
That yowt the heid's wee spinner.
A doughty fechter in the fight,
Wi' posture, and good balance,
An' shak', will she, the hand o' God,
When life's brief flash is darkened.

Our bairns, and bairns's bairns are fair,
A trace we'll leave behind us,
Meantime the bond o'blood is there,
That joins and keeps and twines us.

A fortun'd, fair starred man am I,
Health's, wealth beyond conception,
And love thy neighbour as thy self,
Precludes the deil's pre-emption.

Contents

Foreword

At first glance, the detailed study of how equilibrium thermodynamics may be used to analyse fuel cells seems far removed from the mainstream of fuel cell development. Fuel cell developers are necessarily caught up in a trade where increased irreversibilities are embraced in return for reductions in capital cost and comparisons are made with heat engines from the flawed but familiar mindset of energy and heating values. However, such perspectives, comfortably close though they are to current thinking, ignore not only the full potential of fuel cells to impact the exchange of useful work but also the benefit to be gained from a fresh approach to the engineering of fuel cells from a fundamental perspective.

The study of seemingly abstract aspects of quantum mechanics and solid state physics led paradoxically not only to semiconductor devices themselves but also to fresh insights into how pragmatism could be used to engineer these devices into even more useful forms that could be connected in a highly integrated fashion and dramatically reduced in size. Fresh study of the detailed thermodynamics of fuel cells is at the very least a valuable intellectual exercise for any practitioner in a fuel-cell-related discipline. With a little historical reflection, few could ignore that it might lead to further insights that could directly benefit the pragmatic development of commercially saleable fuel cells.

In Mr Barclay's latest work we see not only a more comprehensive equilibrium thermodynamics analysis than in his previous works but also a much awaited bridge to the practical world of fuel cell development along a path where irreversibilities are recognised for what they are – undesirable concessions in which potentially useful work is given up. The crux of Mr Barclay's book is his revision of the nature of isothermal oxidation, the process which generates the potential difference of a fuel cell, at zero-current equilibrium, or when generating current and power. Mr Barclay places the maximum possible work of a fuel cell at a very high new level, and dismisses the calorific value of the fuel as a basis of comparison for fuel cells. The revision is far reaching, and will undoubtedly be accepted only slowly and begrudgingly by a field that is quick to open up fresh searches for paradigm shifts but is slow to recognise brilliant fresh thinking in its midst.

The book is doubly ambitious in attempting to point the way to the 'hydrogen mine' or source of low-cost hydrogen. Moreover it highlights the difference between compressible gaseous fuels and relatively incompressible liquid fuels for fuel cells.

Mulling over of the book will be an extended task for fuel cell and hydrogen economy protagonists, and we must await the results. Perhaps a piece of common ground is to look forward to economical coal utilisation via distillation of gaseous fuel for integrated fuel cells and gas turbine engines. Or then again perhaps, armed with fresh thermodynamic insight, there are better ideas to be had on how to use nuclear power to extend the work obtainable from gas and oil.

Gerry Agnew
Vice-President, Rolls-Royce Fuel Cells
Derby, UK

Introduction, and Commentary on Matters Affecting all Chapters

A child's amang ye, takkin notes!
And, faith, he'll prent it.
                              Robert Burns

This introduction is rather long, since it includes that matter which is common to all of the following chapters in the book, some of which are new and unique to the book. A first-time lead is provided into the revolutionary new technology of isothermal oxidation, detailed in the thermodynamic appendix (Appendix A). The reaction in a fuel cell is isothermal, charge exchange, oxidation. Combustion does not occur, and its theory does not relate to fuel cells, nor does its main parameter the calorific value or combustion enthalpy.

The author's position is roughly that implied in Burns' remark above. The fuel cell community is young and interdisciplinary. Retired fuel cell technologists are still rarities, so that distillations of long-term extensive experience are hard to come by. Moreover, such experience will not have been against the new background fully developed in this book for the first time. The new fuel cell isothermal oxidation theory is remarked upon in the foreword, and was partly introduced by Barclay

Fuel Cells, Engines and Hydrogen – An Exergy Approach   Frederick J. Barclay
© 2006 John Wiley & Sons, Ltd

(2002). It is a noteworthy worldwide failure that isothermal oxidation is not the industry-recognised modus operandi of the fuel cell, although it is without doubt that it should be, and must be in the future, in the author's view.

It is the function of this book to correct this time-hardened situation, and to refocus the priorities of the development problems of the worldwide industry, from fuel cell makers to hydrogen manufacturers. The confusion is such that the fuel cell should be redefined. It is not enough to refer to continuously fed fuel and oxidant electrodes. The interfacial chemical reaction via isothermal charge interchange at the electrolyte–electrode interface is the other essential feature. The fuel is oxidised (electron removal) and the oxidant reduced (electron addition). Any supposed resemblance between reversible isothermal oxidation and irreversible combustion should go into the rubbish bin of history. A hydrogen–oxygen gaseous fuel cell, and the regenerative **incompressible liquid-fuelled** redox flow battery of Chapter 2, share isothermal oxidation as their modus operandi. The redox flow battery cannot be classified as anything other than an alternative kind of fuel cell. Denying that the redox flow battery is a fuel cell is equivalent to King Canute's battle with the tide. See the latter part of point 10 below for additional remarks on the definition of the fuel cell's modus operandi.

1. *Isothermal chemistry in nature.* Nature in its vegetation and its animal life uses isothermal chemistry to achieve high efficiency. The fuel cell is the first human device to follow nature. The industry, however, has largely failed to see it that way.

   Whereas an aeroplane has a Carnot limited engine with a hot exhaust, the familiar, but contrasting, example of the swan is a significant flying machine propelled by isothermal muscle power, rather than an engine. The cells of living creatures are packed with small chemical plants, termed mitochondria. These, nourished by blood, have the task of providing the exact materials to enable the creature to be self-maintaining, and of manufacturing the fuel for contractile muscles. This fuel is named adenosine tri-phosphate, which hydrates to adenosine di-phosphate when used, and is reprocessed to tri-phosphate in the mitochondria. The complex, liquid reagent, isothermal, chemical process of the mitochondria is known as the Krebs cycle (Campbell *et al.*, 2006) but the reversible entropy changes of its near equilibrium thermodynamics are not yet tied down. The swan, however, does give us a prima facie indication of very high efficiency. Swimming on a pond, the swan takes in small animals, bugs and vegetation. From that apparently meagre fuel it can produce the substantial power needed to take off into flight. No combustion engine fired with the latter fuel could achieve such a performance.

Twenty-five years ago in Madras (now Chennai), India, the author witnessed the excavation and spoil removal, for the foundations of a nuclear power plant, by a numerous team of women in scarlet saris. They had bid for and won the contract against bulldozers, a triumph for low-cost, efficient, rice-fuelled muscle power. With the graceful deportment born of their traditional task of head-high water carrying, they moved spoil in shallow circular trays on their heads. They walked up out of the foundation pit in an organised spiral, effectively an inclined plane. The reader should recognise that rice from photosynthesis is the product of another isothermal chemical reaction which absorbs the sun's power, without raising the temperature of delicate greenery. We, who are muscle powered, are therefore also solar powered.

In ancient cases like the construction of the Egyptian pyramids, muscle power did not have a competing mechanical alternative.

The oft quoted example of the isothermal photosynthesis reaction is fructose production. Carbon dioxide and water are forced together, by solar **power**, catalysed by chlorophyll. Oxygen is released. That is,

$$6CO_2 + 6H_2O = C_6H_{12}O_6 + 6O_2$$

2. *Isothermal chemistry in fuel cells.* Barclay (2002) wrote a paper which is seminal to this book, and may be downloaded from the author's listed web site. The text and calculations of this paper are reiterated, and paraphrased, extensively in this introduction. Its equations are used in Appendix A. The paper, via an equilibrium diagram, draws attention to isothermal oxidation. The single equilibrium diagram brings out the fact that a fuel cell and an electrolyser which are the thermodynamic inverse of each other need, relative to existing devices, additional components (concentration cells and semi-permeable membranes), so as to operate at reversible equilibrium, and avoid irreversible diffusion as a gas transport mechanism. The equilibrium fuel cell then turns out to be much more efficient than a normal fuel cell. It has a greatly increased Nernst potential difference. In addition the basis of calculation of efficiency obviously cannot be the calorific value of the

fuel, since irreversible burning is not involved. The correct basis is asserted to be the fuel chemical exergy, as defined by, and calculated using, the equilibrium diagram of Barclay (2002) and in Appendix A of this book, Sections A.2.16 and A.2.17.

3. *Engine efficiency.* There is also the buried and forgotten fact that engine efficiencies as usually calculated enable one engine to be compared with another, but are gross overestimates of absolute or exergetic efficiency. The comparison of fuel cell with engine is greatly affected. This subject is enlarged upon in point 17 below, on combustion irreversibility,

4. *Multidisciplinary fuel cells.* In the fuel cell engine and hydrogen business, communication between the disciplines of equilibrium and irreversible thermodynamics, physical chemistry, electrochemistry, fluid mechanics, materials science and mechanical arrangement, to name but a few, is visibly open to improvement.

   This book, as commissioned the now inactive Professional Engineering Publishing (PEP), was intended merely to be an expansion of the chapter on fuel cells in the author's PEP book on exergy analysis (Barclay, 1998), pp 69–89, and hence to be a review of the status of the competing fuel cell types. However, during the Seventh Grove Symposium, the author grasped the undoubted fact that the chemical reaction in all fuel cells is **not** combustion, but isothermal oxidation, and as a result wrote his 2002 paper. This was clearly incompatible with the fact of isothermal oxidation, which produces power directly **without** the kind of energy conversion occurring in heat engines, to characterise any fuel cell in terms of the calorific value of its fuel, in joules. But that is what the worldwide fuel cell industry was doing, and still is doing. It was, and is, a case of not understanding Joule's experiment, Figure 3.1.

   Incidentally, the author's activities paused after publishing his 2002 paper, while an early version of this book was completed for PEP. Recently, in 2005, the author went through the index of the *Journal of Power Sources* from Vol. 151 in 2005 and back to the Eighth Grove Symposium, Vol. 118, 2003, so as to modernise the text.

   A prime lesson in irreversible process theory is based on Figure 3.1. In that experiment, shaft **power** was dissipated irreversibly by a rotating paddle, to become **energy** in a tank of near

ambient temperature water. The chaotically interactive translation, vibration and rotation of the water molecules is **energy**:

$$1\,W\,s \gg irrev \gg 1\,J$$

or

1 J at ambient temperature cannot be reconverted to 1 W s.

Energy is accessible to generate power only by cyclic processes
(heat cycles) as defined by Carnot. (Carnot cycle theory is outlined
below.) Energy conversion, as in heat cycles, is not encountered in
fuel cells. Fuel cells generate electrical potential difference directly
(Barclay, 2002) from the chemical exergy or Gibb's potential of
their fuel. Potential difference is an amount of work. Hence fuel
cells are not Carnot limited. Practical fuel cells suffer from many
losses or modes of exergy annihilation, or irreversibilities, as exemplified in Figure 3.1. Those modes are the subject of Chapter 3.

A brief digression is that in Einstein's famous equation

$$E = Mc^2$$

the symbol E is in fact for exergy, and not for Carnot limited
energy. Einstein also overlooked Joule's experiment.

There are two incompatible sayings in present-day thermodynamic usage. These are, firstly, that the entropy is always on the
increase (true); and secondly, that energy can neither be created
nor destroyed (false, but rather popular). The former asserts that
there is a continuous supply of energy, and the latter, based on
$1\,W\,s = 1\,J$, rather than $1\,W\,s \gg irrev \gg 1\,J$, denies the existence
of such an energy supply. Energy creation occurs in Figure 3.1.

Gerry Agnew of Rolls-Royce predicts in his foreword to this
book that grudging admission of this efficiency difficulty may take
an extended period. The author is impatient to improve on that
prediction, and presented his work (Barclay, 2002) at an IMechE
conference on 30 November 2005 in pursuit of that objective.
The conference chairman accepted that the calorific value was no
longer a valid performance criterion. In a discussion, the author
presented an extensive calculation of the power yield of isothermal
oxidation, relative to the yield of combustion allied to a heat cycle.

The correct fuel cell characterisation (Barclay, 2002), it is reiterated, is via the 'fuel chemical exergy' in watt seconds, and numerically a much larger quantity than the calorific value in joules. Fuel

chemical exergies are derived numerically for varied cell operating pressures and temperatures, in Appendix A, along the lines initiated in Barclay (2002). Notional isothermal reactant and product circulators (concentration cells) are used in lieu of the irreversible diffusion which moves the gases in real fuel cells. The **greatly enlarged development potential of the complete** fuel cell **with circulators** is pro rata to the large fuel chemical exergy

$$\Delta G = H - h - T(S - s)$$

which, like electrical potential difference, is an amount of work done. Numerically, as calculated in work units, W s, from the equilibrium diagram, Figure A.1, the fuel chemical exergy, ξ_{ch}, is about 2.0 times higher than the calorific value in energy units J (which needs to be multiplied by the cycle efficiency, say 50%, to reach the value of the cyclic output). Note that the existing literature, the actual editions used by the author, including Kotas (1995) and Moran and Shapiro (1993), does not use equilibrium concentrations for fuel chemical exergy calculation (Chase et al., 1985), in the fuel cell chemical reaction, and puts ξ_{ch} numerically equal to the net calorific value. The hydrogen fuel cell (Figures A.1 and A.2), aided by three concentration cells for the isothermal expansion of hydrogen, oxygen and steam, is a fourfold potent device, when correctly calculated, compared with an assessment via combustion theory.

5. *This book versus alternative textbooks.* Four book titles follow, which do not follow the line of Barclay (2002). In any discussion of isothermal oxidation at the fuel cell electrolyte–electrode interface the four example books suffer from a lack of thermodynamic rigour.

   (a) Vielstich *et al.* (2003). This is a large fuel cell compendium with much useful detail but based on irrational combustion theory for performance calculations.

   (b) Hoogers, *et al.* (2002). This book recognises the part played by the change in Gibb's potential but not the significance of chemical equilibrium (Chase *et al.*, 1986), and therefore of concentration cells as circulators.

   (c) Linden and Reddy (2002). This book uses irrational combustion theory.

   (d) Larminie and Dicks (2000). This book uses irrational combustion theory.

There are two other well-known books which also should be commented on:

(a) Kotas (1995). In this book, Figure 2.12 is used to prove that the fuel chemical exergy and the lower calorific value of the fuel, with different units, are numerically equal. The conclusion does not appreciate the need (Atkins, 1995) for the reactants and their product in the van't Hoff equilibrium box to have equilibrium concentrations (dissociation concentrations). The reader must be aware that this affects the whole book and many calculations.

(b) Moran and Shapiro (1993). Figure 13.3 of this book takes no account of the need for equilibrium concentrations in fuel cell reactions. Figure 13.4 also needs a reversible reformer to feed H_2 and CO to separate fuel cells.

This book, however, seeks to correct the apparent 150-year-old distorted perspective of the industry, highlighted by the incompatibilities between, and contradictions amongst, the texts above and the author's published paper (Barclay, 2002).

Sir William Grove's interpretations of 150 years ago, that his fuel cell notion was the inverse of his electrolyser notion, need to be updated before relatively efficient **complete** fuel cells, integrated with comparatively low-performing gas turbines, are developed and can reach the market (Figure A.7). The integration serves the purpose of burning in gas turbine combustion chambers fuel which the isothermal oxidation reaction is unable to consume. In the practical system, the result is superior starting (preheating) capability, by driving the gas turbine compressor from the power grid. Irreversible burning, in a combustion chamber, is unfortunately the only option, but an inefficient one to isothermal oxidation in the fuel cell.

It is again asserted that the numerical basis of efficiency in fuel cells, engines and combinations thereof is the fuel chemical exergy, derived from Figure A.1. The resulting unwelcome and tedious problem **must** be faced of changing mind sets, and a mass of efficiency figures, in the downwards direction. The good news is that the development potential of the fuel cell is bigger pro rata than the large fuel chemical exergy, with the proviso that new perm-selective membranes and concentration cell circulators need to be developed, before the superior potential can be realised.

The combustion engine is stuck with a major irreversibility, and gets no increase in development potential. It can never realise the fuel chemical exergy. See point 16 below.

6. *Liquid and gaseous reactions.* Every reader will be familiar with the torch cell and with its solid consumable electrodes. In contrast, a fuel cell has continuously fed liquid or gaseous electrodes. Gas systems need much greater pumping power than do liquid systems. The liquids are incompressible and require little circulating power, a major technical advantage, which is incorporated in the trademarked power storage system Regenesys. That system was acquired in 2004 from RWE Fuel Cells, Germany, by VRB Power Systems, of Vancouver, Canada. RWE, in turn, invested in VRB, as will be remarked upon further in Chapters 1 and 2 and especially in Appendix A.

7. *New perspectives arising from isothermal oxidation.* The next chapter of this book describes the greatly altered perspective of the fuel cell industry, when Grove's ideas are updated. The second chapter describes the detail of Regenesys, or ESS-RGN. This system has changed hands, as noted above, and information is available from http://www.vrbpower.com/. (The initials VRB stand for Vanadium Redox Battery, a low-power alternative to Regenesys.) The new 2005 VRB Power Systems shorthand is ESS-VRB for 2.5 to 10 MW and ESS-RGN for 10 to 100 MW. In Chapter 2 the reader will be acquainted with ESS-RGN, one of the two VRB fuel cell systems (incompressible liquid based) which can be termed 'complete'. The redox battery uses small pumps as circulators.

In Weibel *et al.* (2005a, b) Harvard University has announced that it is developing a new, incompressible reactants, liquid-based fuel cell for the isothermal oxidation of a coal/water/sulphuric acid slurry, and producing compressible carbon dioxide. Further detail is given at the end of Appendix A.

All the other gaseous reactant fuel cell systems are 'incomplete' in that they do not have circulators, and move reactants and products by irreversible diffusion. Systems such as the proton exchange fuel cell and its companion the direct methanol fuel cell are doubly incomplete, since they lack a 'hydrogen mine' which can produce cheap hydrogen and cheap methanol from natural gas (see Section A.1.4 and Figure A.4, and Barclay, 2002).

Grove did not realise the irreversibility of his diffusion-based fuel cells and electrolysers, without circulators. Barclay (2002) shows that with circulators an isothermal device is possible which is in precise equilibrium at zero current. Drawing a very small current represents perfect fuel cell action. Supplying a very small current represents perfect electrolyser action, a matched pair, from one device. A solution involving two different devices is not possible, since that would preclude reversibility, as supposed by Grove.

Grove's proposal for a hydrogen economy, based on imperfect fuel cells and electrolysers using compressible gases, is therefore doubtful, and has become an unlikely competitor for ESS-RGN.

Fuel cell and electrolyser irreversibilities, and kinetics in general, are treated in Chapter 3.

The proposed Icelandic hydrogen economy is a very special case based on geothermal electricity of zero fuel cost and very low total cost. The economics overween the poor thermodynamics. Even so, the news is of slow progress.

Hopefully the reader will gradually be induced to start reading and using Appendix A, which, with its detailed calculations, elaborates and underscores the foregoing points.

8. *Starting point of this book.* The fuel cell problem initially appeared to be in circular form, since there was no obvious starting point or end point. But the author had, in both editions of his book on exergy analysis (Barclay, 1995; 1998), the equilibrium diagram of a fuel cell, with circulators. The books sold well, but the equilibrium diagram did not. That failure became the starting point of this book, in which it was necessary to work out numerically, and fully for the first time, the maximum yield of the equilibrium diagram electrochemical processes, namely the 'fuel chemical exergy', as partially done in Barclay (2002), which was also a starting point.

9. *Power from the sun.* From the foregoing, it becomes clear that human utilisation of the **power** output of the sun, by combustion (non-isothermal oxidation) of photosynthesised fuel to feed heat engines, is extremely inefficient, relative to muscle power. Relatively efficient isothermal oxidation in complete fuel cells needs to be made practical and economic, and needs to be seen as the equal and opposite analogue of isothermal photosynthesis, as described above.

10. *Exergy.* The precaution is taken, in this introduction, of intro-
    ducing the word exergy as 'the go of things': exergy, **not** energy!
    Fuel cells are not energy converters. They are direct power, or
    exergy, or work potential generators. Energy, exploitable in a
    Carnot limited heat engine, is a property of molecules en masse:
    for example, the inaccessible energy in the ambient air comprises
    chaotic, interactive, molecular translation, rotation and vibration
    energy. Heat is a flow of energy, part of such everyday experience
    as touching a hot pipe. **All energy is a product of exergy anni-
    hilation,** a continuous process since the Big Bang, the inception
    of the universe, from the timeless, exergy-packed, zero-energy,
    absolute zero temperature, 'primaeval atom' conceived in 1925
    by the Belgian priest and physicist Lemaitre (Singh, 2004) as
    containing all the matter in the universe.

    On a much smaller scale, in Joule's experiment, Figure 3.1, the
    energy of the stirred liquid is increased pro rata to the work put
    irreversibly into stirring, as mentioned earlier.

    Fuel cells are not electrochemical combustors (a contradiction
    in terms). Fuel cells directly generate an electrical potential differ-
    ence (work done on a unit charge moving between the electrodes
    via the electric circuit) which, in thermodynamic units, becomes
    the work done by unit mass of fuel, consumed by the anode
    and cathode reactions, and experiencing a change, ΔG, in Gibb's
    potential. Nernst dealt with the change of units in his famous
    equation (Eqn (4.1)). An equilibrium fuel cell generates V_n, the
    Nernst potential difference, while a **complete** fuel cell generates
    additional V_n, across its isothermal concentration cell circulators.
    Moreover, the sum of the entropies of formation of the reac-
    tants is not the same as the entropy of formation of the product.
    Often, there is a reversible entropy increase with an increment of
    oxidation, and a reversible decrease if the oxidation increment
    is reversed. In other words, the equilibrium hydrogen cell has a
    reversible entropy increase. There is an associated reversible heat
    output, which must be clearly distinguished from irreversible heat
    such as that due to electric current passing through an ohmic
    resistance. The latter occurs in a fuel cell away from equilibrium
    and is an irreversible loss (IR drop).

11. *The audience for this book.* Hopefully, even with its relatively
    difficult analysis, this book will benefit the world fuel cell indus-
    try, and provide a logical new theoretical sense of direction for
    its executives, for the numerous new graduate and doctorate

entrants of all disciplines in that expanding industry, and its associated hydrogen manufacturing partners across the world. For the circular arguments, some familiarity with the industry must be assumed, and some acquaintance with, or enthusiasm for, thermodynamics, physical chemistry and electrochemistry. The work in thermodynamics of J. Willard Gibbs is a vital historical ingredient (Gibbs, 1961).

Economic fuel cells are an unachieved target. Lower capital cost is the main current target, but the industry now should recognise its past errors, revise its thermodynamic basis, and tackle a resulting series of major development problems. A book of limited size cannot teach the many disciplines to be met, so further reading must be involved, notably of patents. Pre-reading is necessary of essential references, such as Ralph, *et al.* (1997; 2002; 2003) and Kumar *et al.* (1995).

Experience of having been in real difficulties with immature nuclear power equipment has had a dissolving effect on the author's disciplinary interfaces, a dissolution recommended for staffers of the immature fuel cell industry.

12. *Keeping up to date.* For the vital purpose of keeping up to date in an industry grappling with a new theoretical basis, a major new aide has appeared (under rapid development), namely http://www.scholar.google.com/.

Use of the singular search words 'fuel cell' gives 41 000 references from the past and up to the current year, 2005, able to be sorted in various ways. 'Fuel cells' is a different search, which leads to books. The author is unable to present descriptions of fuel cells which will remain valid for the period of mulling over this book and its thermodynamic appendix, predicted in the foreword. The industry must evolve rapidly if it is to survive and prosper. Some illustrations useful in this book are left as Internet references to accustom the reader to that essential means of acquiring learning.

Another relatively new updating source is http://www.sciencedirect.com/ for journals and books.

Time-honoured sources are http://www.arXiv.org/ and http://www.desy.de/spires/. Moreover, at the web site http://www.fcr.iop.org/ the reader may examine 'The Fuel Cell Review' from the Institute of Physics and IOP Publishing Ltd, Bristol.

Major professional institutions in the UK, USA and Europe have libraries that can be searched online, for example http://www.iop.org.uk/, http://www.imeche.org.uk/ and http://www.iee.org.uk/. The IMechE Library intends, in the near future, to make available online back issues of the *Journal of Power and Energy*, notably Barclay (2002). The latter, however, can be downloaded from the author's web site, http://www.exergysource.com/.

Since much of this book was written on the basis of information from patents, instructions are given for searching out patents, their existence, their texts and their illustrations. The details are given at the end of the book. The status of patents is subject to change during the period in which they evolve from patent applied for, to patent pending, to full permanent patent. Their ownership changes as companies are taken over.

13. *Fundamentals of combustion.* Air for a combustion engine is heated, or in other words, supplied with energy, by the burning of fuel. Collision at the level of two molecules of hydrogen plus one of oxygen results in two high-velocity water molecules which collide and equilibrate with, or may ignite, other molecules in the vicinity – a stochastic, dissipative and fundamentally irreversible chain reaction not subject to ready control.

14. *Contrast with isothermal oxidation.* The process of isothermal oxidation (Marcus, 1964; 1982; see point 19 below) in a fuel cell of the theoretical complete variety can, in contrast, be made reversible and controllable. Figure A.1 shows the equilibrium diagram of an ambient temperature and pressure cell. Irreversible steps of any kind must not occur in such a diagram. Oxygen is separated from the atmosphere by a perm-selective membrane, and passed to the cell via an isothermal concentration cell which generates a Nernst potential difference (see Nernst's equation, (4.1)). The oxygen concentration is reduced to steam dissociation level, as calculated using the fugacity based tables (Chase *et al.*, 1985; 1998). Steam is present at the fuel electrode at atmospheric pressure. Dissociated oxygen reaches the oxygen electrode, but no gases can back diffuse from the cell because of perm-selective membranes. The fuel concentration is handled by the same technique so that it matches stoichiometrically the oxygen concentration. The steam concentration is reduced from atmospheric in the cell to the standard concentration in the atmosphere given by Kotas (1995). The cell is calculated in Appendix A, and the calculations involve the work

of circulators, ΔG isothermal and ΔB isentropic. That is, in the half-flow oxygen isothermal circulator

$$\Delta G = \Delta H - T\Delta S$$

and there is a work output, from concentration reduction. Now $dg = dh - sdt - tds = -RT \ \ln(P_i/P_f)$ for an isothermal reversible expansion in a perfect gas, where the enthalpy is a function exclusively of temperature. In the expansion of a perfect or ideal gas, therefore, the struck-through terms are zero.

For isentropic compression of the oxygen the term $B = H - T_0 S$ is required (Barclay, 1998; Kotas, 1995). An isentropic process has $\Delta B = \Delta H$, and

Oxygen isothermal circulator $\Delta G = T\Delta S$ = power output
(flow = 1/2)

Oxygen isentropic circulator $\Delta B = \Delta H$ = power input
(flow = 1/2)

Hydrogen isothermal circulator $\Delta G = -T\Delta S$ = power output
(flow = 1)

Hydrogen isentropic circulator $\Delta B = \Delta H$ = power input
(flow = 1)

Chemical reaction in cell, $\Delta G = \Delta H - T\Delta S_{rev}$
= electrical power output

Note that the chemical reaction involves an equilibrium, high-vacuum, dissociated mixture and is not at standard conditions. The Carnot cycle feeds on $T\Delta S_{rev}$.

Steam isentropic circulator $\Delta B = \Delta H$ = power output
(flow = 1)

Steam isentropic circulator $\Delta G = \Delta H - T\Delta S$ = power output
(flow = 1)

The above relate to Figure A.2, which shows an enhanced version of Figure A.1, designed to allow operation of the cell at any selected high temperature and pressure. Isentropic circulators are incorporated to generate the increased conditions. The cell generates heat which is passed without temperature difference to a Carnot cycle to generate power, a reversible process free from the irreversibility of combustion.

15. *Carnot cycle.* The Carnot cycle of Figure A.2 has a rectangular temperature–entropy chart, with upper and lower isothermals and left and right isentropics.

A Carnot isothermal is computed by its ΔG, and an isentropic is computed by its ΔB:

$$\Delta G = \Delta H - T\Delta S \quad \text{and} \quad \Delta B = \Delta H$$

$$\Delta G_{12} = H_2 - T_1 S_2 - H_1 + T_1 S_1$$

$$\Delta B_{23} = H_3 - H_2$$

$$\Delta G_{34} = H_4 - T_0 S_4 - H_3 + T_0 S_3$$

$$\Delta B_{41} = H_1 - H_4$$

Summing, the H terms cancel. Moreover, $S_1 = S_4$ and $S_3 = S_2$, and the net result is $(S_4 - S_3)(T_1 - T_0)$. In words, the area of the temperature–entropy diagram gives the power output of the Carnot cycle. Note that the result is independent of perfect gas theory.

16. *Details of isothermal oxidation.* Figures A.1 and A.2 together illustrate isothermal electrochemical reactions (Marcus, 1964; 1982; see point 19 below) which can, distinct from combustion, be slowed down near to, and stopped at, reversible equilibrium. If the author can convince the fuel cell and engine communities, now in an awkwardly entrenched position, not to characterise their devices in terms of the enthalpy of combustion (calorific value), this book will have been worthwhile. Both for fuel cells and engines, and for combinations thereof, the correct characterisation is via 'the fuel chemical exergy', in units of work, not of energy. The archaic term 'Gibb's free energy' is again a contradiction in terms, since it is a work function, with watt second units (W s) in this book. The name would be corrected if it were Gibb's 'exergy' or Gibb's potential.

As noted above, work is numerically equivalent to energy, but the units are different:

$$\text{Work of 1 W s} \gg \text{irrev} \gg \text{1 J of energy}$$

The one-way arrows indicate that the conversion, for example by stirring a liquid, is totally irreversible. Accordingly $1\,\text{J} \neq 1\,\text{W s}$, it being impossible to unstir! Energy, therefore, is not reconvertible to, or equivalent to, work or power (rate of working). Much analysis is spoilt by ignoring the one-way arrows, notably the fashionable analysis of combined heat and power, which involves the addition of heat and power as if they had the same units.

In this book, the potential difference is used to the exclusion of the electromotive force, which is an archaic misnomer, since potential difference is an amount of work and not a force. V_n is used; E_n is discarded.

17. *Combustion irreversibility.* To understand the foregoing points the reader will need to acquire, or have, a clear distinction in mind between heat yielding, chemical exergy destroying, combustion and work/power yielding, exergy conserving, isothermal oxidation in a fuel cell. The combustion reaction destroys chemical exergy, and therefore commences with a substantial loss. It is reiterated that it is a branched chain reaction in a cool milieu, the product molecules being slowed down from very high velocity by repeated collisions. Equilibration, or thermal degradation or exergy annihilation, is inherent and unavoidable. In a carefully designed combustion engine, thermomechanical exergy, a component of the energy left after the initial chain reaction degradation, is utilised to do work. The ratio between the latter energy and work quantities is called the efficiency. But the other, larger and overlooked part is chemical exergy destruction via the slowing down of high-velocity product molecules. Hence the calculation of engine efficiency based on the calorific value of the fuel greatly overestimates the engine efficiency, by ignoring the slowing-down loss. The remedy is to be realistic and to base engine efficiency on the fuel chemical exergy, and hence to lower the efficiency greatly. Doubtless, that will require some of the predicted period mentioned in the foreword.

18. *Dissociation equilibrium and the JANAF tables.* At equilibrium, gaseous products (e.g. steam) exhibit dissociation into the gaseous reactants hydrogen and oxygen, which is a balanced, zero-work, reversible interaction. The calculations and equilibrium diagrams of Appendix A expose the reader to the foregoing concepts. The ability to calculate the dissociation concentrations is essential, and requires Figure A.1 and the fugacity-based **JANAF thermochemical tables** (Chase *et al.*,

1998). Fugacity is relevant to imperfect water substance, but not to perfect carbon dioxide at practical temperatures. At 160 US dollars the tables are a good investment.

In principle the above calculations can be done from first principles using Section A.2.1 and Eqn (A.1) (reproduced below), together with fugacity data from Moran and Shapiro (1976), but that would be laborious in the extreme, and is not recommended:

$$K = (P_{H_{2}0}/P_{\theta})/(P_{H_{2}}/P_{\theta})(P_{O_{2}}/P_{\theta})^{1/2}$$

In reality the P (pressures) should be F (fugacities), excepting only $P = F$ (arbitrarily) at standard pressure P_0 and temperature T_0. The subsequent calculation involves the equation of state for steam/water.

The JANAF tables (Chase *et al.*, 1985; 1998) are introduced below with three examples of equilibrium constants as a function of temperature at $10 \times$ atmospheric pressure, and also at atmospheric pressure, for the oxidation of hydrogen.

Two example tables are given from the set for the dissociation of water substance. These are in contrast to the single table for carbon dioxide which, being a perfect gas at relevant temperatures, contains data independent of pressure, not requiring fugacity in its analysis.

JANAF Example Table 1. The full table (Chase *et al.*, 1985), p. 1276, rises from 0 K to 1800 K in 100 K intervals with irregular stops at 298.15 K and at 453.070 K, the latter being the saturation temperature for 10 bar, at which the table has some twin values for liquid and of the saturation line, and the vaporous end. The reducing log K_f values in the far right column indicate increasing degrees of dissociation of water substance to hydrogen and oxygen with temperature, or increased equilibrium concentrations. Table pressure $= 1.0$ MPa $= 10$ bar.

T(K)	C_p^0	S_0	$-[G^0 - H^0(T_r)]/T$	$H^0 - H^0(T_r)$	$\Delta_f H^0$	$\Delta_f G^0$	$Log\,K_f$
298.15	75.313	69.946	69.946	0	−285.815	−237.125	41.543
500	41.002	186.240	85.919	50.16	−244.579	−209.567	21.903
1000	41.517	213.517	143.626	69.891	−247.955	−173.468	9.061

JANAF Example Table 2. The full table (Chase *et al.*, 1985), p. 1275, is for 1 bar, saturation temperature 298.15 K. Comparison with the first table shows that the values of $\log K_f$ and the resulting equilibrium dissociation concentrations are a function of pressure, denoting the use of fugacity in the analysis of imperfect water substance.

T (K)	C_p^0	S_0	$-[G^0 - H^0(T_r)]/T$	$H^0 - H^0(T_r)$	$\Delta_f H^0$	$\Delta_f G^0$	$Log\, K_f$
298.15	75.351	69.95	69.95	0	−285.83	−237.141	41.546
500	35.699	206.428	104.712	50.858	−243.896	−219.069	22.886
1000	41.292	232.730	162.736	69.994	−247.868	−192.593	10.060

The product from hydrogen and oxygen reactants may be steam/water, or superheated steam, very imperfect or non-ideal fluids. Only one equilibrium can exist at a time. In other words, two different equilibria cannot co-exist, and give rise to perpetual motion. The product water substance is dissociated into rarefied hydrogen and oxygen, dictating the low concentrations at the fuel and oxygen electrodes. For accurate fuel cell calculations, involving imperfect H_2O, the use of fugacity is a requirement, eased by the calculations in the table above, given fully in Chase *et al.* (1998). However, if the fuel is carbon monoxide, the product is carbon dioxide, a near ideal set of gaseous reactants and product, then fugacity need not be used. An idealised equilibrium fuel cell has a pressure coefficient of zero when fuelled by carbon monoxide, but a negative pressure coefficient when using hydrogen, and when calculated using the fugacity-based JANAF tables. Away from equilibrium new variables arise, and the pressure coefficient is likely to be negative.

19. *The works of R. A. Marcus.* (See the Internet for his list of 341 publications, and the total is rising.) There is no quantum chemistry in this book, but the towering, quantum-based, very extensive writings of Nobel Prizewinner R. A. Marcus, (1964; 1982) are the source, via the two quoted examples, of a qualitative understanding of redox reactions between electrode electrons and vibrating hydrated ions (H_3O) in the potential gradient of the electrolyte double layer, or equivalent phenomena in non-aqueous high-temperature fuel cells. Depending on the phase of the ion vibration, electrons are either emitted to the electrode

gas from the electrolyte ions or emitted by the electrode gas to the electrolyte ions. The reversible, equilibrium, alternating to and fro action has been termed the exchange current. (More correctly exchange flux, since the electron trajectories are not confined to being normal to the interface. They may cross it in any direction.)

The advent of scanning tunnelling microscopes, atomic force microscopes and electrochemical microscopes (quantum-physics-based machines) has made familiar the existence of a thin layer of tunnelling electrons, 'clouds' immediately outside the conducting porous electrode/catalyst structure and inside the non-conducting electrolyte. As a minimum, such electrons must influence the statistics of equilibrium near to the electrodes. Balanced reversible **exchange currents**, unique to each electrode, are the consequence of equilibrium redox reactions which generate a dancing, advancing and retiring population of additional electrons in the electrolyte solution. In the absence of tunnelling electrons from the electrode structure, there would be a chicken and egg problem: that is, the source of a stray electron to start up the exchange current reactions.

20. *Vector analysis.* The author learnt vector analysis in a 1950 postgraduate course, based on the German book of 1932, *Classical Electricity and Magnetism*, by M. Abraham and R. Becker, published by Blackie and Sons, Glasgow. The book was written so that its definitions of scalar flux (flow in all directions) and of vector current (flow in one direction) fitted both hydrodynamics and electromagnetism. The same definitions were carried over unmodified into the scalar neutron fluxes and vector neutron currents of the nuclear power reactor. In electrochemistry, however, the term 'exchange current' attaches to what is more properly described (see above) as a local, somewhat anisotropic, scalar flux.

The underlying, or associated, clouds of tunnelling electrons in the non-conducting electrolyte, clearly an anisotropic scalar flux, got the name 'jellium' from the electrochemist Schmickler (1993). The consolation is that the mathematics is often right, even if the language is difficult, or idiosyncratic. The electrochemist's system has to be joined, not beaten!

When speculating about the uncharted aspects (Ralph *et al.*, 1997; 2002; 2003) of electrocatalysis, say, with platinum crystallites having adsorbed surface hydrogen atoms and having surface dislocations, tunnelling electrons must be a factor to consider: Schmickler's 'jellium' (Schmickler, 1993). Indeed, the contemporary use of layers of joint electron/ion conducting interface materials on cathodes productively **extends** the 'jellium' notion. The theory of proton conduction in aqueous electrolytes (Koryta, 1991; 1993) depends on proton tunnelling between polar water molecules which happen to be aligned correctly and take time to rotate into favourable alignment.

The foregoing quantum phenomena are dealt with in 'further reading' references, notably Marcus (1982), see point 19 above. It is a fact that electrochemical reactions (and therefore exchange currents) concentrate at surface platinum dislocations with associated hydrogen atom irregularities. The mechanisms of electrocatalysis are obscure. Precisely how does platinum so strongly influence the electrode exchange currents at equilibrium? The activation 'energy' (a misnomer for exergy) for **reversible** electron transfer at cell equilibrium must be zero. Hence the popular hearsay which asserts that platinum reduces the activation energy for electron transfer is certainly incorrect at equilibrium. It may have some validity away from equilibrium, when the balanced exchange currents are submerged by the unidirectional cell current common to both electrodes!

21. *Dislocations and catalysis.* The dislocations per unit platinum area may be a significant number, not yet measured or referred to in the literature! The reversible hydrogen electrode may owe its performance to a high concentration of dislocations on the surface of platinum black.

22. *The fuel cell industry.* The fuel cell industry has changed rapidly while the author caught up on it during the writing of this book, in which the descriptions of practical, air-breathing fuel cells can be no more than ephemeral snapshots of immature (incomplete) devices, without circulators. Nevertheless, the relationships between the ideas of Appendix A and the incomplete and therefore fleeting practical systems of the remaining chapters are educational. Numerous web sites appear in a list given in Appendix C, so that the reader will be able to follow the future victories and setbacks of the firms in the industry, notably against the background described in this introduction and in greater detail in Appendix A.

The quicksand of economics is a dangerous area, where a change in interest rates or fuel costs can demolish what was a reasoned design. Robust economics may be achievable by considering, and by minimising, sensitivity to arbitrary economic fluctuations. Fuel cells are not yet at that stage, they are still expensive and, as this book asserts, poorly understood. But they have environmental, thermodynamic and economic promise. Niche markets are visible, for example stationary non-interruptible power supplies, based on generous fuel storage. The high value of operational continuity has the result that expensive, inefficient, incomplete fuel cells are nevertheless competitive in such specialised applications.

Currently, the stationary power market penetration of the fuel cell is based on reduced local pollution rather than superior performance. Indeed the internationally demonstrated Ballard fuel cell bus generates more pollution from the power plant stack which generates its hydrogen supply from an inefficient incomplete electrolyser, than it saves by emitting steam from its exhaust. The industry has to rescue itself from this untenable position.

Cell performance calculations, based on the calorific value, are shown to be irrational above. Hence the industry lacks a generally accepted logical and technical perspective, and it is the author's assertion that only exergy analysis will suffice. But whereas the industry thinks of such analysis as difficult and unrewarding, in reality the rewards are crucial to the future. The difficulty must be vanquished.

23. *Remarks on present-day incomplete cell technology.* The natural-gas-fuelled solid oxide fuel cell (SOFC), as currently proposed for small stationary power plants, comprises a reformer in series with a fuel cell. Integration with the fuel cell anode is not practical at high Westinghouse SOFC temperatures, but works well at lower MCFC (Molten Carbonate Fuel Cell) temperatures, or certain intermediate IT/SOFC temperatures, when an inherently irreversible reform reaction is coupled to a near reversible Faraday cell reaction. The irreversibility of the electrode reformer using electrode reversible heat is less than that of the separately fired reformer, because the latter includes the irreversibility of the combustion reaction. Just visible in the laboratory, and of high promise for the future, is the elimination of the reformer by direct hydrocarbon electro-oxidation (Perry-Murray *et al.*, 1999; Parh *et al.*, 1999; Gorte *et al.*, 2000).

24. *Reversible reform.* The endothermic reform reaction itself involves a heated catalyst, heated reactants, including excess steam, and cool products: namely, much irreversibility. In principle, reform could be carried out (see Figure A.4) in a Faradaic, electrical work consuming, apparatus arranged to feed a pair of H_2 and CO fuel cells. If that is done on paper, without assurance of practicality, then the yield with methane fuel is the chemical exergy of methane, just as if the methane had been directly oxidised, as above. The latter represents an alternate calculation path to the fuel chemical exergy. Both reformer and direct oxidation routes cross-check in Appendix A, with similar quantitative results. The minimal reversible ΔS, accompanying isothermal oxidation of methane, could lead to electricity production with minimal heat (Gardiner, 1996), but since there is always significant unused fuel in a fuel cell exhaust, to be rescued only by combustion of the surplus, the latter ideal recedes without prospect of return.

25. *Patents.* With this book, one writing difficulty has been the unwillingness of fuel cell vendors to display their hardware, sometimes because of perceived immaturity of their patents. But their patents have provided a route to **non-copyright** black and white illustrations, gratefully acknowledged, which would have been the only product of seeing the hardware.

The desk top computer, via the Internet and such US compilations as delphion.com, or the various national patent office compilations, notably www.patent.gov.uk/, esp@cenet and www.uspto.gov/patft/index.html/, provides access to patents, greatly aided in the author's case by the British Library paper and electronic systems. Patents provide useful literature references and access to the inventors' basic thinking, right or wrong! Patents, the author has found, are full of mistakes, inconsistencies and contradictions, which enable the reader to learn by comparison, and to sort out the quintessential information. See Appendix C, for an elementary patent search technique via the US Patent Office.

26. *Systems to be described.* One main objective of the book has evolved to become the presentation of a limited, rather than comprehensive, list of incomplete fuel cells, as available at the time of writing. There is small value in an omnivorous approach, because no present-day fuel cell without circulators has long-term

survival capability. Another objective is to clarify the need for complete fuel cells, modelled on equilibrium diagrams.

The placing into administration of ZEVCO, the Zero Emission Vehicle Company, in the latter part of 2001, encouraged the author not to write a chapter on the incomplete alkaline fuel cell. However, there is a transatlantic revival of this incomplete system at present (end of 2004) which the author has ignored as ephemeral.

Historic references to Bacon's cell are given in Chapter 1. In the case of the sulphuric acid cell some UK work at Cambridge University and ENECO extends the original efforts of Shell in the 1960s. All such systems are 'incomplete' and the author is not moved to include such ephemeral items in his limited review.

Similarly, the phosphoric acid fuel cell (PAFC), although still selling, is resisting efforts to get its capital costs down to the level at which it could be mass produced, and hence another chapter is eliminated. Reference is, however, made to the part played by the PAFC in evolving the catalysts for the polymer electrolyte fuel cell (PEFC).

Chapter 2 appears on the ESS-RGN of VRB Power Systems, Vancouver, formerly the Regenesys power storage system, both for its own sake and for the thermodynamic points it enables the author to make, reiterate or emphasise in relation to his thermodynamic theories. Here are fuel cells without oxygen, and certainly without combustion, but undoubtedly with charge transfer reactions separated by a potential difference. Moreover, here are fuel cells with incompressible liquid reactants and products.

Still in the race are the PEFC and its stable-mates the direct methanol fuel cell, the molten carbonate fuel cell and the solid oxide fuel cell. Although new candidates for the contest are still appearing, the author will ignore them until one appears which has considered the need for circulators, or as at point 7 above (Weibel *et al.* 2005a, b), a new cell appears with incompressible reactants (but not incompressible products). The race is interactive with the choice of fuel infrastructure, and the author points out that hydrogen manufacture from natural gas, as by Methanex or Air Products, is in principle amenable to greatly reduced irreversibility, by paying attention to the calculation route of Figure A.4. Such an improvement would highlight the low efficiency/great cost of electricity from heat cycles, based on irreversible combustion. Ultimately for stationary power, a principal remedy is via complete or fully developed SOFCs with circulators, for direct oxidation of natural gas, and integrated with gas turbines. Alternatively coal can be distilled to provide a gaseous fuel supply for complete fuel cells integrated with gas turbines. The day of primacy in power generation for fire and the heat cycle must be nearing its end. With the advent of an outline of a route to the hydrogen mine involving simultaneous production of hydrogen and power, the possibility appears of efficient methane utilisation in vehicles and the chance that potentially dangerous ocean ecosystem acidification by carbon dioxide absorption can be slowed down to counter the effects of the growing economies of the USA, Europe, India and China and others. A new danger is the reflection of solar heat by atmospheric particulates, which may disappear as a result of engine improvements, and so maximise any carbon dioxide greenhouse effect. The greenhouse effect by carbon dioxide is a controversial topic, but ocean acidification is a certainty of unknown danger to the ocean ecosystem. The experiment was undertaken with no prior consideration or capability for reversal.

27. *Electrical equipment.* Fuel cell power in the form of direct current at low potential difference will usually need to be adapted to practical use, by conversion to alternating current at, say, 110 V, 60 Hz or 230 V, 50 Hz. Two main inverter equipment suppliers have their web sites listed at the end of the book, namely the Asea Brown Boveri Group, suppliers to Ballard Power, which has most of the fuel cell market, and Xantrex Technology Inc. Ballard is just entering the market, see Ecostar below.

A major design objective of any inverter is the supply of sinusoidal voltage waveform, free from troublesome harmonics (Power Conversion Section, 2005a, b; 2003). The steady DC potential of the fuel cell must be interrupted with precise regularity to correspond to 50 or 60 Hz output, or the desired variable frequency for some vehicles. The resulting waveform is likely to be a function of the interruption technique. One slogan is hard switching versus soft switching. Improved waveform can be achieved by filtration using inductance and condenser combinations, to be optimised in vehicle applications for low weight. The timing of the interruption enables the supply of both active (watts) and reactive power (vars) to achieve good voltage regulation.

The Renault Scenic fuel cell vehicle (see the listed web site) has a specially developed high-efficiency squirrel cage induction motor for each road wheel, and is likely to have rivals in competing projects. However, at http://www.sustainability. renault.com/html/image.htm?img=images/p/p.jpg, Renault shows a car layout with a front axle driven by one motor. There is a 48 V DC to DC converter followed by a DC to AC inverter. The hydrogen source is a petrol reformer. Evidently the project is still fluid. Setting out the options and alternatives is in hand.

The scale of the fuel cell projects with ABB power converter/inverter equipment varies from 1 kW mains-coupled domestic power and heat installations to 100 MW Regenesys power storage plants supporting the activities of power grids. The Regenesys electrical equipment is not as large as its power rating suggests, since only liquid pumps are involved rather than gas isothermal circulators. The details are dealt with at the end of Chapter 2, via the ABB article 'Electricity from the Store'.

The literature is full of obscure mnemonics, to be dealt with piecemeal as they occur.

The simplest project is the fixed indoor fuel cell, which can for example provide a guaranteed power source against the chance of mains failure.

Available from abb.com is the pamphlet 'ABB Power Conditioning Systems, DC/AC Converter (2003) for PEM Fuel Cell System Hydrogenics, Mississauga, Canada, 80 kW Indoor Fuel Cell Converter'. The selected mains conditions are 60 Hz, 575 V. The reader has to interpret this in 'Inverter Control: AMC Control Cards' and in 'Inverter Technology: IGBT, Voltage Source Inverters', and then in 'Communication: Modbus Serial Link'. The DC to AC conversion is performed by an ABB voltage source IGBT (Insulated Gate Bipolar Transistor) inverter unit. This voltage source inverter is capable of controlling power from the fuel cell stack to the grid while compensating for the stack DC voltage which varies as a function of power output. The unit also has an additional control feature allowing system self-support with power from the fuel cell if utility power is lost. The inverter modules use forced air cooling with a fan. Three air-cooled sinus filters assure harmonic currents and voltages within IEEE 519 standards. A dry-type VPI power transformer provides galvanic isolation and voltage matching between the fuel cell inverter voltage and the 575 V grid voltage. Grid and load voltage digital monitoring, inverter and fuel cell current control, inverter charging and sequence control, protection and monitoring are all implemented in the enigmatic 'digital RMIO' controls. A simple key pad with display provides a local human–machine interface, and a digital serial interface is used as a control interface with customer control systems.

Also obtainable from abb.com is PCS 500 – Power Converter Systems' modular standard converter system, showing PEBBs (Power Electronics Building Blocks) a modular approach to converter design.

Ballard Power developed in 2004 the Ecostar power converter. The system is not yet extensively applied, but is clearly a sign of future competition for ABB and total independence of the competition for Ballard. See ballard.com.

In Gopinath *et al.* (2004) the development of a prototype 12 kW fuel cell inverter at the Texas Agricultural and Mechanical University is described. In order to read the document on the Net, one has to join the IEEE. However, by going to the British Library the author was able to bypass that difficulty, and obtain a printout. This document, selected from 12 recent documents, does not explain its own mnemonics, not even DSP. It is worth reading nevertheless, because the scheme visible via text and circuit diagrams is clearly very different from the ABB inverter for Hydrogenics Mississauga, above. The first stage is a 48 V to 400 V direct voltage converter, followed by a DC/AC inverter outputting 120/240 V at 60 Hz AC. The switch devices are IGBTs, as in ABB practice. No remarks are made on the intended application, but there is a battery backup, and such vehicle accessories as supercapacitors are mentioned. Moreover, the fuel cell voltage of 48 V is indicative of the latest vehicle practice, but the power level of 10 kW is not comparable with the 78 kW of the Honda FCX. See newspress.co.uk. A fuel cell substitute for the lead–acid battery of a luxury car may be the application to be guessed.

A second IEEE report, which does explain its mnemonics, and which competes for mention here, is by Tolbert *et al.* (2002), which predicts the arrangement of the electrical equipment for heavy hybrid electric vehicles capable of operating in fuel-cell-only mode, engine-only mode and with both power sources. To achieve minimum THD (Total Harmonic Distortion) use is made of cascade multilevel voltage inverters allied to batteries and ultra-capacitors. It then has to be contrived that the 11-level converter scheme has to be rearranged automatically each day to ensure even battery charging, an impressive scheme.

The site http://www.dpi.wa.gov.au/fuelcells/technology.html contains useful pictures and information on the equipment of the Mercedes Benz Citaro fuel cell bus, an adaptation of a standard engine-driven bus. That information can be added to by the scholar.google reference 'Demonstration of Fuel Cell Buses under Varying Climate Conditions – The Stockholm CUTE Project'. See the Internet for Fuel Cell Bus Club. Powered by Ballard PEMFCs, and fuelled by pressurised hydrogen, the fuel cell bus has a single rear axle drive from a single 200 kW motor. The bus has DC to DC conversion followed by DC to three-phase AC inversion. If the hydrogen supply is from an electrolyser, then it has green attributes if the power is hydroelectric as in Stockholm. If the power is from a thermal power plant, then as this book points out, the electrolyser will consume more power than the bus generates, so that the plant stack will create more pollution than the bus saves. The pollution does get moved to a better position for dispersal, however.

28. *List of symbols.* For compactness in the calculations in the text, units are intermittently shown, but always comply with this list.

u = Specific internal energy, $kJ\,mol^{-1}$
h = Specific enthalpy, $kJ\,mol^{-1}$
v = Specific volume, $m^3\,mol^{-1}$
g = Specific Gibb's potential, $kW\,s\,mol^{-1}$
s = Second, of time
t = Temperature, °C or K
p = Pressure, bar
m = Mass flow, $kg\,s^{-1}$
z = Amount of fuel in a gas mixture, e.g. z_{CO} = kg of carbon monoxide
P_0 = Standard atmospheric pressure, 1 bar
PP_0 = Partial Pressure of component of standard atmosphere, bar
K = Absolute temperature in kelvin
°C = Temperature in degrees centigrade
T_0 = 25 °C = 298.15 K = Temperature of standard atmosphere
S = Specific entropy, $kJ\,mol^{-1}\,K^{-1}$
C_p = Specific heat at constant pressure = $(\partial u/\partial t)_p$

C_v = Specific heat at constant volume = $(\partial u/\partial t)_v$

ΔC_t = Carnot cycle power, kW

ΔC_r = Circulator power, kW

ΔG = Fuel cell power, kW

γ = Specific heat ratio = C_p/C_v

γ_{CO} = Fraction of carbon monoxide in an equilibrium mixture of gases

ξ = Specific exergy, $kW\,s\,mol^{-1}$

$\Delta\xi_{ch}$ = Change of chemical exergy = fuel chemical exergy = work of standard conditions isothermal oxidation in an equilibrium fuel cell, with circulators, $kW\,s\,mol^{-1}$

$\Delta\xi_{tm}$ = Change of thermomechanical exergy = work of a reversible expansion or compression process, or of a reversible heat cycle, $kW\,s\,mol^{-1}$

1

Altered Perspectives

The man who makes no mistakes, does not usually make anything.
Edward John Phelps, 1822–1900

In what follows we are building on the work of Sir William Groves, who made much, but committed one major sin of omission.

The introduction in this book of the equilibrium isothermal oxidation theory of **complete** fuel cells with isothermal circulators (concentration cells) imposes on the industry changes of perspective relative to existing non-equilibrium, Groves-type, **incomplete** fuel cells without circulators, and dependent on gas movement by irreversible diffusion in a concentration gradient. Equilibrium isothermal oxidation is in stark contrast to fundamentally irreversible combustion. The irreversibility of combustion is an overlooked, neglected, phenomenon, which the reader must from hereon retain in mind. The high-speed products of combustion are slowed down in the combustion gas, their speed being non-recoverable. They heat the gas and cool themselves in an irreversible process. A large irreversible entropy growth occurs. The kinetic exergy of the high-speed product molecules is destroyed. This destruction is ignored in traditional engine analysis, to produce efficiencies which are far too high, but do enable each engine to be compared with any other.

The modified perspectives in the sections below relate to fuel cells (liquid and gaseous), engines, fuel cells integrated with engines, hydrogen sources and the hydrogen economy.

Fuel Cells, Engines and Hydrogen – An Exergy Approach   Frederick J. Barclay
© 2006 John Wiley & Sons, Ltd

Would tae God, the Giftie Gie Us,
tae see Oor'sels, as Ithers see us!
                    Robert Burns

This second quotation seeks a refreshed sense of proportion about things, exactly the mission of the chapter to come.

In the author's introduction, the history of this book is explained, and initial stage-setting technical remarks are made. The history is expanded below, and new topics added, which make clearer the very large effect on the fuel cell industry of the new isothermal oxidation thermodynamic treatment using the equilibrium diagram, Figure A.1. The figure is the basis for detailed calculations of the fuel chemical exergy, in Appendix A.

As is common sense, the results of the performance calculations are dimensionless numbers, W/W, for actual power/ideal power. That cannot be said for calculations based on the calorific value, which are in units of $W J^{-1}$. The use of 'fuel chemical exergy' in W s, a large number compared with 'calorific value' in joules, as the basis of fuel cell and engine efficiency calculations, greatly lowers the numerical efficiency of both systems, a realistic event. The unavoidable lesson is that all existing means of power generation based on heat cycles are woefully inefficient. Only complete fuel cells offer high efficiency, and that in the future after some problem solving.

The fuel cell development potential, when the necessary circulators and membranes are added, becomes very much larger than that of the engine, a greatly encouraging realisation and a spur to intense future development to improve upon the low efficiencies of the last sentence.

Comprehension of the difference between compressible gases as reactants and incompressible liquids gives an illuminating new sense of proportion about the overweening relative merit of power storage by redox battery in ESS-RGN versus the older, doubtful, Groves proposal for a hydrogen economy – doubtful because Groves uses inefficient and incomplete electrolysers and fuel cells dependent on irreversible gas diffusion for gaseous reactant and product movement.

1.1 POWER STORAGE

The one and only fuel cell being manufactured, which makes use of circulators, as in Figure A.1, is the power storage system ESS-RGN, formerly Regensys, which features liquid reactants which are incompressible in contrast to the gaseous reactants of all other fuel cell systems.

The circulators for liquid reactants are mere pumps, which elevate the liquid level in storage tanks. A full description of Regenesys, now owned by VRB Power Systems of Vancouver, Canada, is given in Chapter 2 so that the reader can compare and contrast this 'complete' system with the gaseous and 'incomplete' systems of later chapters. From the terminology, it is doubtful if VRB itself has grasped the underlying reason for the superiority of its two power storage systems. VRB speaks of energy storage, when (based on Joule's experiment), it should really speak of power storage. VRB is silent on the subject of incompressible liquid reactants and products.

1.2   CIRCULATORS

Figures A.1 and A.2, and the equilibrium calculations of Appendix A, lead to huge pressure ratios for the required isothermal concentration cell circulators. These are not existing, developed devices, but are ideas made necessary by Figures A.1 and A.2. The incorporation of fully developed circulators in a practical non-equilibrium gaseous cell, distinct from Figure A.1, would involve their operation at a reduced, more practical pressure ratio, relative to the operating point of the cell on its V/I characteristic, Figure 6.5, rather than relative to the zero-current point of equilibrium at the top of the figure. The initial steep slope of the figure results from readjustment from zero-flow equilibrium at the zero point to irreversible gas diffusion at the operating point. The latter explanation is unique to the author.

Mechanical circulators are probably not usable as they require intercoolers/heaters to achieve isothermal performance, and have a very limited pressure ratio.

1.3   INCOMPLETENESS

Incompleteness in Chapters 4–6 is of two kinds, lack of circulators and lack of a 'hydrogen mine', or cheap source of hydrogen. Both of these represent major development problems, as do semi-permeable or permselective membranes, which must accompany circulators and feature in the flow sheet of the hydrogen mine, Figure A.4.

The ICI Billingham plant (Barclay, 1998), and its more modern successors operated by Methanex and by Air Products, provide expensive hydrogen from natural gas reform via combustion heat. The elimination of combustion-heated reformers is needed, since they involve large combustion irreversibilities. Such plants are optimised using the technique of 'process integration', which gives the best arrangement of irreversible heat exchanger networks, having a sequence of temperature changes.

1.4   THE HYDROGEN MINE   *page* 151.

The author asserts that an isothermal plant arrangement (process integration therefore not applicable) is essential for greater thermodynamic efficiency, and hopefully for cheapness. An equilibrium diagram for such a plant is shown in Figure A.4. The plant is arranged so that exergy is supplied as electrical power and not by combustion heat, an entirely theoretical concept, and therefore a difficult development problem, for Methanex, Air Products and Ballard.

An efficient plant must be based on making an approach, as near as is practical and economic, to the perfect processes of an equilibrium diagram, such as Figure A.4, which was initially composed as a calculation route for methane chemical exergy, and then realised to have larger implications. Methane had to be consumed in an isothermal equilibrium reversible process.

The process of the calculation involves a reformer which gets its exergy not from combustion, but as electrical power generated by supplying CO and H_2 to separate fuel cells which are able to create an excess above the need of the reformer. The invaluable JANAF thermochemical tables (Chase *et al.*, 1985) provided the thermodynamic data. The excess is the chemical exergy of methane.

Viewing Figure A.4 as a potential hydrogen mine, the change would be made that the excess hydrogen would be stored, and converted to power in separate stationary or mobile fuel cell power plants rather than in the fuel cells of the diagram. Such a hydrogen mine is both a major development problem (involving circulators, fuel cells with some fuel in their exhausts, as well as a new kind of reformer) and a pressing need. The hydrogen mine **must** somehow be realised to conserve the large fuel cell development and international investments of Ballard Power in the fuel cell vehicle. Note that the internal, fuel-cell-based power generation of the hydrogen mine would be very much more efficient than any combustion-based heat cycle. The way forward is for all non-nuclear power generation to be based on complete fuel cells with circulators and integrated with gas turbines.

Air Products, in its web site question and answer session, goes over the current status of hydrogen production by reforming natural gas, but predicts no future process change. But Fuel Cell Energy in its web site News Releases & Events discusses the prospect of a hydrogen energy station to produce power, heat and hydrogen simultaneously, in conjunction with Air Products: 'The HES will integrate aspects of FCE Direct Fuel Cell (MCFC) Technology, with Air Products' process and separation technology'. The HES sounds akin to Figure A.4, but no comparable flow sheet is given.

The Methanex Vancouver web site lays emphasis on methanol production and hopes for the direct methanol fuel cell (DMFC). Hydrogen is mentioned, but not the need for a hydrogen mine for the nearby Ballard Power Inc.

The hydrogen industry clearly needs to refresh its sense of proportion and begin to tackle its major new development problems. Likewise the fuel cell industry has much to reappraise and many new problems to tackle.

> And those behind cried forward, and those in front cried back!
> Baron MacAulay, 1800–1859

At such a juncture as the one covered here, the best way forward is not easy to establish.

1.5 COAL GASIFICATION

A large question, looked at below, is the future of gasified coal, supplying complete fuel cells, integrated with gas turbines and with greatly reduced carbon dioxide output, relative to present-day plants based only on heat cycles.

Coal gasification (Hotchkiss, 2003) to provide clean fuel for gas turbine power plants is an established competitive modern technology, with Spanish, UK and US examples. The fuel gas produced is largely carbon monoxide, but with a fraction of hydrogen. Enough coal is burnt to heat the remaining bulk of coal in a distiller closed to the atmosphere, and of such design that solid debris does not reach the gas turbine inlets.

The flow sheet of a Hotchkiss gasifier is given in Figure 1.1.

A future development would be to pass the fuel gases to the circulators of high-temperature complete SOFCs or MCFCs, integrated with gas turbines, that would result in a major gain in fuel efficiency. The

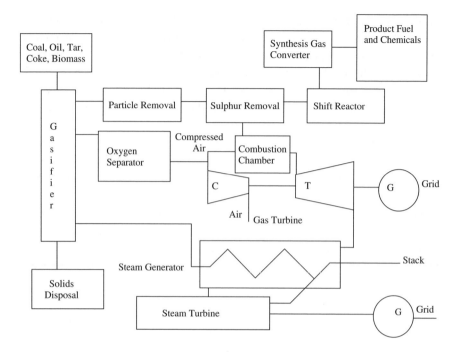

Figure 1.1   Coal distillation gasifier flow sheet

operating point of the fuel cells and the distiller operating point of the fuel cells would be interactive design variables.

In India and China, plentiful and cheap coal seems likely to be used to supply the power demands of these major populations. From the angle of the ocean acidification problem, and of the more complex greenhouse gas problem, a large economy in fuel consumption, by using fuel cells allied to coal gasification, would be a welcome new development. Both countries are also pursuing vigorous nuclear power programmes, which will, once operating, produce no carbon dioxide.

1.6   SOFCs

The designers of SOFCs have to face the new developments posed by the circulators of Figures A.1 and A.2. The letters SOFC denote a vigorous, diverse and expanding family of fuel cells based on the idea that at high temperature the thermal oscillations of the ions at the electrolyte/reactant/product interfaces lead to vigorous exchange current at equilibrium, and to usefully large ion conductivity at the selected operating point, some way from equilibrium. With the commonly selected electrolyte of yttrium-stabilised zirconia (YSZ), an operating temperature of 1000 °C lead to good ion conductivity.

However, 1000 °C leads to a very rapid reaction if anode reform is attempted and in many cases the result is excessive thermal stress of the ceramic electrolyte, so that conventional reformers must be used. As a consequence there has come about a class of intermediate-temperature SOFCs based on alternative ceramic formulations, 500 °C operation having been achieved by a metal/ceramic fuel cell by the company 'Ceres' (see Chapter 4) set up by Imperial College London.

The nascent ability of some SOFC versions to oxidise methane directly (Perry Murray et al., 1999; Park et al., 1999; Gorte et al., 2000) using appropriate catalysts represents a major challenge to the rival MCFC, which cannot emulate the new technology. The direct isothermal oxidation of methane as in Figure A.5 means that the new system has no need for a 'hydrogen mine', although the need remains as an essential for vehicle fuel cells.

The SOFC exists in rectangular and circular form with flat plate membrane electrode assemblies (MEAs) and in tubular form with fat tubes and small tubes, the whole range being described in Chapter 4. An important variable is the thickness of the electrolyte layer, which is a manufacturing problem involving porosity if the layer is too thin. Losses are minimised if the layer is thin. The specialised small-tube SOFC must have relatively thick walls for tube strength, and sacrifices performance to achieve resistance to thermal shock and thermal stress due to anode reform. The Rolls-Royce design of the 50 W all-ceramic rectangular plate fuel cell, Section 4.3, was aimed from conception at mass production, and via ingeniously compact arrangement at large stacks for multi-megawatt power production. Gas turbine integration was and is intended.

An attempt by the author to forecast the design of a plant based on complete IT/SOFCs with concentration cell circulators, integrated with a gas turbine and consuming natural gas, is shown as Figure A.7.

1.7 MCFCs

The USA, Germany and Japan are large players in the business of the **incomplete** MCFC. Fuel Cell Energy, USA, its licensee MTU Friedrichshafn, and the Mitsubishi Materials Company all have to face up to the irreversibility caused by the lack of circulators. Details are in Chapter 5.

A strength of the system is its ability to undertake anode reform, with a comfortable reaction rate, and low thermal stress at 600 °C. The ability of the MCFC to achieve 40 000 h operating life is still being demonstrated by its protagonists. Moreover, changes to the molten carbonate formulation are under consideration. Observed mobility of the molten carbonate, within the matrix conceived as fixing it, has been ascribed by the author to the surface tension gradient associated with concentration and temperature differences.

The MCFC is suitable for integration with gas turbines, and Japanese studies of this topic are extensive. But there are no prototypes as for the SOFC (Siemens Westinghouse). Two figures taken from Chapters 4 and 5 are reproduced below for the reader's convenience. The first figure (Figure 1.2) is the Siemens Westinghouse SOFC/gas turbine integration scheme.

The second Figure (Figure 1.3) shows the author's thinking on MCFC/gas turbine Integration.

In brief the MCFC is somewhat less mature than its pressing competitor, the SOFC. The design of MTU Friedrichshafn is impressively compact and ingenious.

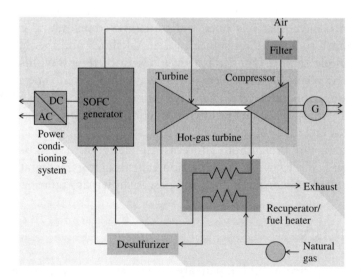

Figure 1.2   SOFC/gas turbine integration scheme

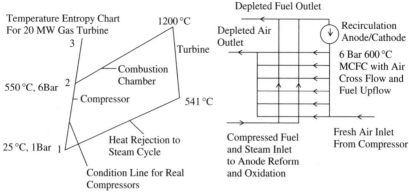

Notes

1] The temperature entropy chart is that of an existing 20 MW gas turbine.

2] An ideal machine would have a compressor outlet temperature equal to the MCFC, namely 600 °C.

3] The mismatch could be overcome by burning a little fuel at the combustion chamber inlet, and supplying the fuel cell with slightly depleted air, at 600 °C. Better to select, for simplicity, a matched gas turbinel.

4] The fuel cell would be large, way beyond anything currently contemplated.

5] There are no Published Test Data for MCFC installations. There are some mathematical models.

Figure 1.3   MCFC/gas turbine integration scheme

1.8   THE PEFC

The Ballard PEFC has been and is being developed in a well co-ordinated way. The basis is international. The MEAs are made in the UK by Johnson Matthey. The low-cost electrode structures/flow plates are formed from flexible graphite (Grafcell), from the US supplier Advanced Energy Products (formerly Graftech), but with catalyst platinum particles from Johnson Matthey Two lower cost alternatives to the classical proton exchange electrolyte, DuPont Nafion 112, are being developed in the UK by Victrex Ltd. The alternatives are Ballard Advanced Materials BAM3G and a Victrex material, Peek. See Chapter 6. These new materials are not available to competitors, who will have to purchase licences or do their own development as have the Japanese and Chinese.

In Xuezhong Du *et al.* (2001), Japanese 'Aciplex' and 'Flemion' and Chinese 'Shanghai' alternatives to Nafion are compared with Nafion in terms of their performance, and none is overwhelming. The latter alternatives have not yet caught on in the international scene. Price comparison is not made available.

Fuel cell stacks are assembled by Ballard and tested in application situations such as vehicles and stationary power. Operating experience is fed back. Bus demonstrations in many large cities displace the pollution from the vehicle to power plant stack. And not only displace but increase, since the hydrogen is produced in inefficient electrolysers without circulators.

The PEFC system itself will have to face the consequences of irreversibility due to lack of circulators. Crucially, the system must have a 'hydrogen mine'. Perhaps Ballard will need to join with Methanex and Air Products to solve this major development and capital investment problem. There are no signs of that happening.

In order to minimise the sensitivity of the platinum catalyst to CO and CO_2 impurity in reformed fuel gas there has been a rise in PEFC operating conditions to 150 °C and corresponding pressure. A reason for favouring such a change is the response of the system to extreme climatic conditions such as 100% relative humidity, which the author has encountered in the Middle East, at 38 °C. Getting rid of steam water product in such conditions would be made easier by the operating temperature of 150 °C, compared with the usual 70 °C. Note that the idealised systems of Figures A.1 and A.2 exhaust to arbitrary standard 20% relative humidity. All extreme conditions need actual tests, and Ballard has done the tests on altitude by operating vehicles at 7000 ft (2130 m) in Mexico City. Reduced oxygen concentration does have an adverse effect, but not a disastrous one.

1.9   ENGINES

At the time of writing of the UK steam tables in the 1930s by Professor Callendar, he and other authorities on thermodynamics decided that widespread and thorough understanding of the work of Gibbs (see Gibbs, 1961) by engineers was unlikely. Such understanding has now become essential and indispensable to rational power production.

Accordingly, a formula was devised which enables comparison of the performance of one heat engine with another. The air flow through the combustion chamber of a gas turbine, for example, is heated by the experimentally determined calorific value of the fuel flow. That gives a heat input, to be compared with shaft power, to produce an efficiency. Dimensional analysis indicates, however, that such an efficiency is not a pure number, as an efficiency should be. The shaft output has to be compared with the chemical exergy of the fuel input, the latter being the maximum power theoretically available, calculated from Figure A.1. The exergetic efficiency so determined is very much lower than the conventional efficiency. Reiterating from above, the loss ignored by the conventional efficiency is the equilibration of the high-speed product molecules with their combustion air environment, a huge irreversibility.

The conventional efficiency has served the industry well, but cannot live on into a fuel cell era dominated by fuel cells integrated with engines (see Figure A.7). The fuel cell and the engine must share an identical efficiency basis. The integrated engine, operating in a low-efficiency cyclic way alongside a fuel cell stack, operating via efficient steady flow isothermal oxidation, must be seen as providing a marginal, but useful, increment of performance to the joint, fuel-cell-dominated, power production system. The gas turbine also provides an extremely useful stop/start capability, by the operation of the gas turbine and its compressor on mains power, to get the system up to ion conduction temperature and the commencement of isothermal oxidation.

2

Regenerative Fuel Cells or Redox Flow Batteries

My father feeds his flocks; a frugal Swain,
Whose constant cares were to increase his store.

John Home

2.1 INTRODUCTION TO THE REGENESYS SYSTEM

This chapter was written during the multiple interregnum of National Power, Innogy, Regenesys Technologies Ltd, RWE Fuel Cells and VRB Power Systems, when the system changed hands frequently and finally settled with VRB Power Systems of Vancouver, Canada. No new information on the Regenesys system has been published by VRB since it became owners of the system, and called it ESS-RGN. From this author's viewpoint the name should be PSS-RGN in accordance with the correct interpretation of Joule's experiment, and the need to achieve clarity by always differentiating between power (PSS) and energy (ESS).

The need for frugality in power supply parallels that in the above quotation on agriculture. Power storage, in an economic form, has been a long time appearing. But it has now appeared in superior form. A description of power storage by regenerative fuel cell is included at this point, not only because of its potential economic importance, but also because the equilibrium diagram, Figure A.1 is of greatly reduced

Fuel Cells, Engines and Hydrogen – An Exergy Approach   Frederick J. Barclay
© 2006 John Wiley & Sons, Ltd

significance when incompressible liquid reactants are used, as in the Regenesys system. Hence the Regenesys process for power storage has much higher efficiency than could be achieved by the processes of the proposed alternative hydrogen economy, based on gaseous reactants propelled by diffusion in irreversible electrolyser/fuel cell pairs, without circulators/concentration cells or perm-selective membranes. A source of losses in the kinetics of Regenesys is the need for ions to diffuse between the anolyte and catholyte chambers, via a Nafion membrane.

The black and white diagrams in this chapter were inspired by the references mentioned towards the end of the chapter, and drawn by the author using the Microsoft Word program 'Paint'. The manufacturer's diagrams were decompressed from their web site (passed into history), and were added during revision of the chapter. They are gratefully acknowledged as being of Regenesys Technologies Ltd origin, albeit the firm has disappeared into RWE Fuel Cells, and then into VRB Power Systems.

The Regenesys system underwent a double change of ownership. It was acquired by RWE as part of a large takeover of UK enterprises, and devolved to RWE Fuel Cells, who shut down the major UK and US prototype rigs and reinvested in conventional 'incomplete' fuel cells based, the author suspects, on erroneous reasoning, namely the calculation of excessive efficiencies based on the calorific value of fuel. VRB acquired the system along with the research director of Regenesys. VRB has operated at the lower end of the power storage market, and may have much of the data from the large rigs. The company owns the Vanadium Redox Battery, (25 kW–10 MW, ESS-VRB), similar in objectives to Regenesys (10–100 MW, ESS-RGN).

The reader, even when reassured by the foreword, may still be reluctant, after reading Chapter 1, to forsake the ingrained notion that irreversible combustion is somehow relevant to equilibrium fuel cells. But here in ESS-RGN are fuel cells, with liquid reactants and products, and without oxygen! Combustion cannot be relevant to the Regenesys system, as described below, and in the abb.com reference (Power Conversion Section, 2004) 'Electricity from the Store'.

In electrochemistry and its electrode reactions, the addition of electrons constitutes reduction; the removal of electrons constitutes oxidation, an accepted generalisation which explains the allocation of the term 'fuel cell' to Regenesys.

With ambient temperature aqueous reactants and products, which are liquids of very small compressibility compared with gases, there is insignificant pump work associated with reactant circulation. In the light of Chapter 1, the latter appears to be a major advantage relative to power storage employing hydrogen and oxygen. The special case of Iceland is based on uniquely low-cost geothermal electrical power and easily available reserves of hydroelectric power. The low power cost outweighs the poor thermodynamics of their electrolyser/fuel cell combination, which is explained below and in Chapter 1.

2.2 HISTORY AND PATENTS

Bartolozzi (1989) summarises work from the 1970s up to 1989 by NASA and its subcontractors on hydrogen/oxygen regenerative fuel cells, which had no outcome comparable with Regenesys as illustrated in Figure 2.1. Here, the function of the pumps is merely to raise and lower the liquid levels in the tanks of bromide catholoyte and sulphide anolyte.

The patents in Sections 2.3–2.10 (Zito and Kunz, 1972; Zito, 1996; 1997; Cooley and Nix, 1998; Duncan *et al.*, 2000; Morrissey, 2001) are the basis for onward and continuing successful power storage development, by National Power Ltd UK, by its offshoot, Innogy Ltd, and finally by a second offshoot, Regenesys Technologies Ltd. The German utility RWE has bought all the foregoing. The Innogy work led to the construction, for black start purposes, at National Power's

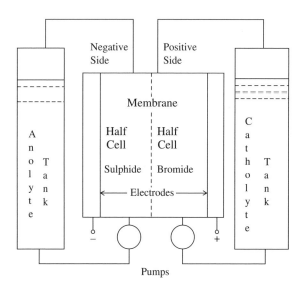

Figure 2.1 Redox flow battery or regenerative fuel cell

Little Barford combined cycle gas/steam turbine power station, of a 15 MW/120 MW h, demonstration redox fuel battery. The installation is described in an abb.com illustrated leaflet, 'Little Barford Power Station, Regenesys-Plant, United Kingdom'. See also Section 2.10 below. A parallel installation proceeded at a TVA site in Columbus, Mississippi. The Little Barford demonstration was preceded by successful experience with a smaller Regenesys prototype at Aberthaw. The work was hedged about with commercial secrecy, but enough has been released, notably in patents, to enable the author to make his teaching points. The above rigs were purchased and terminated by RWE, whilst the Regenesys system was purchased in 2004 by VRB. RWE, perhaps realising a mistake, then bought shares in VRB.

2.3   REGENESYS TECHNOLOGIES LTD; POWER STORAGE

The system is now owned by VRB, under the name ESS-RGN (Regenesys), and illustrated in the basic flow sheet of Figure 2.1. The former Innogy literature refers to the storage of an alias of chemical exergy, namely **chemical potential energy**, a more precise term than the archaic misnomer, 'free energy'. Potential energy (exergy) is of course a work potential, such as the height of a dam!

The efficiency of such a system is simply the kW h power output divided by the kW h power input (above 50% for Regenesys, a figure not approachable in the proposed hydrogen economy via gaseous hydrogen/oxygen, as explained in Chapter 1).

Electrical power (watts), as alternating current, enters for storage and is converted to direct current. Stored power from charged electrolytes is regurgitated as direct current and reconverted to alternating current. The conversion apparatus can generate or consume reactive power (vars), for such purposes as voltage regulation. The system is shown in Figure 2.1. There is a coloured version on the listed vrbpower.com web site.

2.4   ELEMENTARY CHEMISTRY

The Regenesys patents above were explored on the Internet using the Intellectual Property Network, and with the search words 'National Power' and 'Ralph Zito' as obtained verbally at a Grove Symposium.

Sulphur dissolves in solutions of soluble sulphides
and forms a mixture of polysulphide ions, with chain
lengths, 2–5.

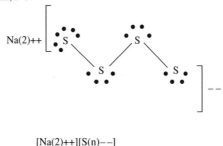

[Na(2)++][S(n)– –]

Sodium Poly-4-sulphide, Aqueous, n = 2 to 5.

The atoms of a polysulphide ion are joined in chains
coupled by single co-valent bonds.

Figure 2.2   Chemistry of sodium polysulphide (after Mortimer, 1975)

Those patents are also listed in the references at the end of this book.
Innogy selected the positive and negative electrolytes of sodium bromide
('anolyte') and sodium polysulphide ('catholyte'). Sodium polysulphide
is shown in Figure 2.2, as a Lewis structure, as in the textbook
of chemistry (Mortimer, 1975). Further reading on polysulphides is
available in Porterfield (1993), Lessner (1986; 1988) and Murray
(1983).

Regenesys uses DuPont's Nafion (Section 6.1.7) as the perm-selective
sodium ion transfer membrane, separating the two half cells, Figure 2.1.
Diffusion of sodium ions in the concentration difference across the
Nafion membrane is one of the irreversibilities of the system. The low-
cost plastic (e.g. polyethylene) tanks and pipework are treated with
fluorine to provide bromine resistance, and are able to operate with,
and contain, both electrolytes at ambient temperature.

Cells with a direct working potential difference of around 1.54 V are
arranged in series modules, Figure 2.3.

The battery is designed to provide the desired voltage for an inverter,
which generates, or takes in, alternating current. Variation of the firing
angle of the inverter (ABB manufacture) gives reactive power (vars) of
either sign, in addition to an input or output of real power (watts). ABB
inverters are further dealt with at the end of the introduction, under
'Electrical equipment'.

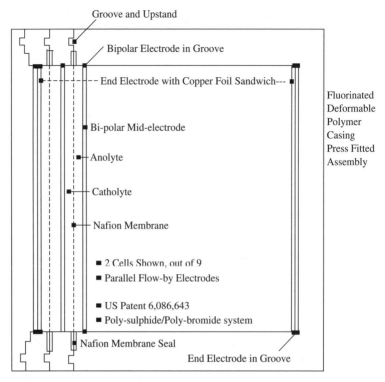

Nine Fuel Cells with Central Nafion Membrane – Regenesys System.

Electrical Power Storage by Fuel Cell, with Liquid Reagents.

Figure 2.3   Power storage in redox battery with liquid reagents

2.5   MODUS OPERANDI OF REGENESYS

During the power storage regime, each electrolyte is pumped through its half cell and stored in one of the two tanks, in 'charged' form (Figures 2.1 and 2.2). During power recovery, the electrolytes are pumped back through their respective half cells and 'discharged'. The +ve cell half reaction, sodium bromide reduction, or electron addition, with $\nu = 2$, is

$$3\{(Na^+).(Br^-)(Aq)\} - 2Na^+(Aq) \text{ (to membrane)}$$

$$+ 2e^- \text{ (from power circuit)} >> \text{charge} >> (Na).(Br_3)(Aq) \quad (2.1)$$

The ionically bonded sodium bromide receives two electrons, and loses two +ve sodium cations to form sodium tri-bromide (see Porterfield, 1993). An alternative homogeneous reaction route to tri-bromide is to bubble bromine (Br_2) through aqueous sodium bromide. Accordingly, the electrochemically formed tri-bromide has an associated atmosphere of bromine from homogeneous dissociation. Such bromine is corrosive, but is contained by the fluorinated plastic tanks and by one side of the (Na^+) Nafion perm-selective membrane. Nafion is itself a compound with many fluorine atoms. Bromine is very much lower than fluorine in chemical reactivity, so that the bromine containment strategy is complete, and the selection of the bromide electrolyte seen to be a well-considered choice.

The polysulphides can be prepared in aqueous sulphide solution, by adding solid sulphur. Being aqueous, the solution contains not only sulphide and polysulphides, but S^{2-} and HS^- ions. Containment again includes the side of the Nafion membrane, opposite the bromide. Migration of HS^- and S^{2-} ions and in the opposite direction Br^2 through the Nafion occurs slowly, with the sulphur moving the more quickly of the two. This migration has resulted in proposals for future systems using multiple membranes and idler electrolytes, to give sulphur recovery and reinjection into the polysulphide electrolyte (Zito, 1997). See below however, for successful operating experience with the simpler, basic system.

The $-ve$ cell half reaction, polysulphide oxidation or electron removal, with $\nu = 2$ as in Eqn (2.1), is illustrated simply as

$$[Na_2][4S] - 2e^- \text{(to power circuit)}$$
$$+ 2Na^+ \text{(from membrane)} >> 2[Na][2S] \qquad (2.2)$$

But clearly, although the poly-4-sulphide may predominate, the other polymers will form an equilibrium mixture (Giggenbach, 1974) in the storage tank.

The illustrative aqueous overall cell reaction showing polymer rearrangement is

$$3NaBr + Na_2S_4 \Leftrightarrow 2Na_2S_2 + NaBr_3 \qquad (2.3)$$

The reaction proceeds, with charge or power input, from left to right. As there is no valency change between Na_2S_4 and Na_2S_2, the polymerised sulphur is referred to as 'zero-valent'.

2.6   SOME CONSTRUCTION DETAILS

The electrode construction is a 'composite consisting of a bromine resistant polymer, and various carbonaceous materials'. Remarks in the patents listed at the end of the book give further information on electrode construction.

'Flow-through' and 'Flow-by' electrodes are considered for differing circumstances. Positive and negative electrodes may differ. The end electrodes of a battery (terminals) are of special construction. The electrodes must be conductive, mechanically stable and highly resistant to bromine diffusion.

Each cell has a single Nafion membrane, and the electrodes are of the flow-by variety. The hot pressed electrodes are stated in Zito (1996) to be based on graphite grains coated with active carbon. After pressing, the electrodes are force cooled for flatness. Bromine impermeability is based on a sealant (Zito, 1997), polyvynylidene fluoride, alternatively polyethylene, in the gaps between the electrically contiguous activated graphite grains.

It may be of interest to record that the development of Nafion substitutes by the Victrex Company, UK for Ballard could be a way of cheapening the system, subject to a licence being granted by Ballard. Additionally, in Xuezhong Du *et al.* (2001) it is recorded that Nafion substitutes have been developed in Japan and China and a technical comparison is made.

2.7   ION AND ELECTRON TRANSFER

The Regenesys system stores the combined ΔG of the two half reactions, and releases the same ΔG for power delivery, with losses of around 25% due to overvoltages, ohmic power dissipation and ion diffusion. Electron transfer and sodium ion movement are the agents of the process:

$$\text{Overall} \quad Br_2 + S^{2-} \Leftrightarrow 2Br^- + S \tag{2.4}$$

$$\text{Half} \quad Br_2 + 2e^- \Leftrightarrow 2Br^- \tag{2.5}$$

$$\text{Half} \quad S^{2-} \Leftrightarrow 2e^- + S \tag{2.6}$$

The above reactions are at ambient temperature, so that an associated Carnot cycle (Chapter 1) is not needed, as would have been the case at high temperature. As the electrolytes are repeatedly reused, the question of the economic or thermodynamic cost of their manufacture is of greatly diminished importance. Nevertheless the cheaper electrolytes have been selected. The system is in the realm of the electrochemist and the chemical engineer, and at the service of the power engineer.

2.8 POWER STORAGE APPLICATIONS

Power storage has wide possibilities, both civilian and military. Submarines with power storage could sprint without engine overload, or proceed stealthily with minimum noise and with engine propulsion on standby. The output of tidal or wind power systems could be steadied, adding value. Islands can smooth their power consumption and minimise transmission peaks. All sorts of vehicles could have their engine output steadied, to the benefit of fuel consumption. A major competitive sorting process will determine which applications of power storage are economic. As noted above, there is development potential visible in the shape of double or treble ion selective membranes with an 'idling electrolyte' in the resulting compartments, enabling sulphur to be removed from the bromine side and returned to the sulphide side (Zito, 1996). These multiple membranes would complicate the layout of Figure 2.3, but see Morrissey (2001).

2.9 INITIAL OPERATING EXPERIENCE

Much information is absent from the preceding description. Development experience is missing, as are economic data. These would become available to customers. However, the flavour of initial operating experience, with beneficial surprises, is given by Morrissey (2001), which is worth detailed reading. The latter reference describes how ultra violet/vibrating ion spectroscopy (UV/VIS) with an attenuated total reflection (ATR) probe may be used to observe, during charge/discharge cycles, 230 nm (and other) absorbance levels, due to sulphide S^{2-} ions, and thereby the state of charge of the redox fuel cell (RFC) within the range 0.5 to 2.5. The state of charge is the ratio of the total number of sulphur atoms in the sulphur/polysulphide electrolyte to the total number of units (charge on an electron) of negative charge carried by all sulphur species present. Simplistically, with states of charge from 0.5 to 2.5, the sulphur species goes from S_2^{2-} to S_5^{2-}. Experience shows that the sulphur equilibrium is not so simple, and as a result there have been beneficial reductions of overvoltage. The resulting cell operating cycle is

the subject of the patent application (Morrisey, 2001). The initial practice was to avoid state of charge values above 1.7, to keep well clear of
the onset of colloidal sulphur, followed by solid sulphur precipitation at
2.2. But the more evolved practice is to enter the state of charge region,
1.7 to 2.5, in which the overvoltages are reduced and the efficiency
increased.

Moreover, sulphur and bromine ion migration across the membrane
is reduced, thereby slowing down the approach to the condition 'unbalanced electrolytes', involving an increased concentration of sulphate ion
and reduced bromine. Examples, with detailed operating results, are
given by Morrissey (2001), in which two operating regimes are compared, 1.3 with 1.8 and 1.3 with 2.15. Results from the latter regime
are better than the former. Details are given of UV/VIS results at four
wavelengths and of sulphur concentration changes.

2.10 ELECTRICAL EQUIPMENT

Power Conversion Section (2004) is an article from the ABB Review,
entitled 'Electricity from the Store'. The article should be viewed and is
contemporaneous with this chapter. It gives plant diagrams not given
in the author's description above, showing, to scale, electrolyte storage
tanks, Regenesys modules, power conversion system, control room and
transformer. As in the electrical systems of the introduction, a vital
component is the IGBT of (Power Conversion Section, 2005a). A circuit
diagram shows the Regenesys modules supplying direct voltage to a
chopper via an inductor in each lead. The chopper comprises two pairs
of four IGBTs each paralleled by a smoothing capacitor. A pair of
rectifiers feeds an inverter comprising two pairs of four IGBTs generating
alternating current to the primary of a grid transformer.

2.11 REMARKS

The author has been fortunate to work for Regenesys during the period
of ownership change. A present-day author would find the path to the
author's sources difficult to identify, because of the demise of significant
UK organisations, and their web sites, along with the dispersion of their
personnel.

In a climate where intermittent power sources from wind, wave and
tide are being built, power storage will be important to the economics
of grid systems. The enforced intermittent operation of nominally base

load plant in natural gas, coal and nuclear power plants would be expensive, giving power storage an economic function. This book focuses on major fuel economies via hydrogen mines and complete fuel cells, but all fuel economies such as the above are vital to minimising firstly carbon dioxide ocean acidification, and secondly climate change in so far as it is carbon dioxide based, rather than being due to solar fluctuations, or solar reflection into space by particulates.

2.12 CONCLUSIONS

The Regenesys power storage system has been described at length, and its operating principles given. Its main loss, or kinetics feature, is ion diffusion through a Nafion membrane separating anolyte and catholyte chambers. Its outstanding advantage is the fact that it uses incompressible reagents/products, and therefore requires minimal auxiliary power. It is sometimes argued that the phrase fuel cell does not apply to this power storage system. There is no actual photosynthesis-originated fuel. However, there is electron addition (reduction) and subtraction (oxidation) at separate potentials, and of course the anolyte and catholyte are continuous electrodes, an essential fuel cell feature. The analogy is helpful to fuel cell system analysts.

3

Irreversible Thermodynamics

A good man, now, nowadays is hard to find!

Bessie Smith

The most famous blues singer of them all had a void to fill. In the fuel cell industry the author finds all fuel cells to be incomplete and imperfect, with only the power storage system of the last chapter being an approach to completeness. This chapter goes into detail on the shortfalls of present-day fuel cell systems.

3.1 CELLS AND ELECTROLYSERS WITH AND WITHOUT CIRCULATORS

Without circulators fuel cells are not the opposites of electrolysers, as supposed by Sir William Grove. Both are unable to operate reversibly between the atmospheric concentration of, say, oxygen and its concentration at the electrode (corresponding to dissociation).

A present-day fuel cell depends for its oxidant kinetics on irreversible diffusion for the transport of oxygen, and sacrifices the potential difference generated by the oxygen circulator, as in Figures A.1 and A.2.

In the case of an electrolyser without circulators, there has to be an oxygen flow to the atmosphere, so that the electrolysis must supply oxygen at much higher than equilibrium concentration which can diffuse to

Fuel Cells, Engines and Hydrogen – An Exergy Approach   Frederick J. Barclay
© 2006 John Wiley & Sons, Ltd

the atmosphere. Only the fitting of a circulator can provide reversible kinetics, and avoid the diffusion irreversibility with its inevitable additional power consumption.

The preceding sentences indicate a situation which has been misunderstood by the fuel cell industry throughout its 150 years of existence, since Grove. A fuel cell without circulators generates reduced Nernst potential difference, and the reduction is the measure of its irreversibility. An electrolyser without circulators, on the other hand, is in need of additional potential difference to produce oxygen at higher than equilibrium concentration able to diffuse to destination. Only when both cell and electrolyser are fitted with circulators can they be equal and opposite, as originally anticipated by Grove for the purposes of the hydrogen economy. That concept, based on gaseous cells and electrolysers, has been overtaken (Chapter 2) by Regenesys, which uses incompressible liquid reactants, and has superior kinetics.

The performance of the apparatus in the isothermal enclosure of the equilibrium diagram, Figure A.1, which may be a fuel cell or an electrolyser depending on which way the equilibrium is tilted, is $1.23V_n$, $237.1\Delta G$. That equal and opposite performance could not be achieved by a PEFC with irreversible chemistry at its cathode as discussed below.

A cell electrolyser pair to operate the hydrogen economy would need simple electrode chemistry. In practice both halves of the pair would have operating points well clear of equilibrium at significant current and reduced voltage. The circulators needed for efficient performance would be matched to the chosen operating points. As noted previously, the losses involved in gas circulation make it likely that liquid-based Regenesys would be an overwhelming competitor for the hydrogen economy. The rational efficiency of any type of power plant is

Actual power output/rate of consumption of fuel chemical exergy

The efficiency of equilibrium isothermal plants as shown in Figures A.1 and A.2 is 100%. Such plants are reversible and free of losses, because of the use of charge exchange processes.

3.2 IRREVERSIBILITY – AN INTRODUCTION VIA JOULE'S EXPERIMENT

Imperfections, irreversibilities or losses are always encountered in real apparatus. They must be seen against the background of a perfect apparatus, such as in Figures A.1 and A.2.

A prime lesson in irreversible process theory is based on Figure 3.1, illustrating Joule's experiment. In that experiment, shaft power was dissipated irreversibly by a rotating paddle, to become energy in a tank of near ambient temperature water. The chaotically interactive translation, vibration and rotation of fluid molecules is energy. Energy is accessible to generate power, only by cyclic processes (heat cycles) as defined by Carnot. (Carnot cycle theory is outlined in Chapter 1.)

Energy conversion, as in heat cycles, is not encountered in fuel cells. Fuel cells generate electrical potential difference directly (Chapter 1) from the chemical exergy of their fuel. Potential difference is an amount of work. Hence fuel cells are not Carnot limited. Fuel cells suffer from many modes of exergy annihilation, or irreversibility, as exemplified in Figure 3.1 Those modes are the subject of this chapter. A brief digression is that in Einstein's famous equation, $E = Mc^2$, the symbol E is correctly for exergy, and not for Carnot limited energy.

Figure 3.2 is shown as a diametrical opposite of Figure 3.1. The shaft power is delivered to/received from a heat pump/heat engine, shown as a reversible process, on a temperature–entropy diagram, relative to the steam/water saturation line. The output of the steam heat pump could be completely recovered by reversing the heat pump to become an engine! If the temperature rise of the heat pump is made very small the delivery of joules tends towards infinity. The delivery of joules reduces with increasing temperature rise. Hence 1 W s can produce any desired amount of heat, but always the same amount of exergy. The temperature–entropy diagram for a steam heat pump is an inversion of

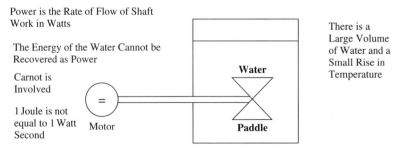

1 Watt Second >> irrev >> 1 Joule

Energy in Joules is the Summation of the Vibration, Translation, and Rotation of Fluid Molecules

Power is the Rate of Flow of Shaft Work in Watts

The Energy of the Water Cannot be Recovered as Power

Carnot is Involved

1 Joule is not equal to 1 Watt Second

Motor

Water

Paddle

There is a Large Volume of Water and a Small Rise in Temperature

The Exergy Flowing along the Shaft is said to be Annihilated; the Carnot Efficiency is Zero

Figure 3.1 Joule's experiment, irreversible transformation, power > energy

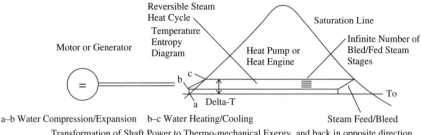

Figure 3.2   Exergy transformation

a regenerative steam heat cycle with an infinite number of bled steam feed heaters. Some detail is shown in Figure 3.2.

The following introductory remarks require the use of many terms of the trade, not previously introduced. Reliance is placed on the reader having some familiarity with the fuel cell business, backed up in this book by information sources such as web sites, references and the index.

3.3   PEFC IRREVERSIBILITY

In Tseung (1997), a British fuel cell pioneer records his frustration of the 1960s upon encountering the experimental fact of hydrogen-fuelled PEFCs with an open-circuit potential difference 18% below the theoretical one! That was a brutal introduction to an irreversibility, Figure 6.2, which, to this day, has not been eliminated, reduced or quantitatively explained. However, there has been success in reducing the amount of expensive platinum catalyst needed for high-current economic performance (Ralph *et al.*, 1997). This shortfall of potential difference between the hydrogen anode and oxygen cathode at open circuit is underlain by the mechanics of mixed potential (Bockris and Srinivasan, 1969), involving parasitic chemical reactions additional to the main intended cell reaction, which occur predominantly at the cathode/electrolyte interface, and result in local circulating currents. These reactions dictate that the open-circuit PEFC does not get close to equilibrium, as envisaged in Chapter 1. The modern PEFC electrode structure (Ralph *et al.*, 1997) has a multicrystalline three-dimensional layer of platinum particles, intended to be in local electrical contact with the gas transmitting porous/fibrous carbonaceous material of the electrode structure, and also in local contact with the Nafion electrolyte membrane. Particles in the membrane, but not in electrical contact with the carbon, catalyse normal non-electrochemical oxidation of hydrogen fuel to water, to produce heat rather than electrical work (Masahiro *et al.*, 1998). The particles are heterogeneous, and catalyse most strongly at dislocation sites, producing local concentrations and concentration gradients with associated irreversibilities. In spite of these zero-current disadvantages, which diminish with increasing current, the PEFC has promising economics based on high power density, flexible graphite and apparent development potential. The PEFC is certainly a very competitive system for vehicles, and is backed by major investors. The hydrogen infrastructure is, however, such as to supply expensive hydrogen fuel, see the introduction.

Some types of PEFC employ bipolar steel plates with carefully machined gas grooves to convey a uniform (or desired) gas flow per unit area to the reaction zone. The current of electrons from the tops of the upstands or groove walls into the same area of the reaction zone is influenced by surface films. The same comments apply to titanium flow plates. The optimisation of this situation is, at the same time, the minimisation of a local irreversibility, namely electrical resistance.

If, at ambient temperature, the slow growth of a crystal is attempted, then, against a background of thermal agitation, the systematic assembly of a perfect crystal turns out to be impossible. Dislocations and interstitial atoms are unavoidable. The second law is obeyed, in that entropy growth is unavoidable. Hence fuel cell materials are real and imperfect, and will need careful optimisation.

A perfectly pure material would corrode. It would have a voracious, concentration-difference-driven hunger for impurities, and would be difficult to employ in practice.

Precisely symmetrical models of aqueous electrode interfaces are often shown for educational purposes. Polar water molecules and positive ions are assembled in a repetitive array. In reality the array would be interrupted by impurities and irregularities, all thermally agitated. The precise achievement of a potential difference V_n, corresponding to Nernst's equation (4.1), would require microscopically uniform conditions at the interface between electrode and electrolyte, so as to avoid local variations of potential difference and circulating currents in the electrode.

Such perfection is aided by assuming, in equilibrium calculations, a pure water electrolyte. The small self-ionisation of pure water is sufficient at zero-current equilibrium. Practical features such as strong electrolytes can be added as perturbations.

3.4   BACON'S FUEL CELL; AVOIDANCE OF IRREVERSIBILITY

Note that Bacon's bi-porous electrode material (corrosion-free, lithiated nickel oxide) is very close to being homogeneous, and enables the achievement of an unmixed potential difference close to V_n. An irreversibility was minimised when the material was selected, on grounds of corrosion resistance, or in other words, to eliminate parasitic chemical reactions involving peroxides of hydrogen (Chapter 6). The latter feature, however, did not, on its own, lead to overall success, via low overvoltages, high power density and superior economics. Having achieved a tidy status at equilibrium, the Bacon cell nevertheless did not achieve competitive performance, at power, except in spacecraft.

3.5   FUEL CELL ENGINEERING

Fuel cell engineering involves economic approximation to the ideal. However, the restricted task in this chapter is to highlight and explain the imperfections of fuel cells and their output shortfalls, relative to the ideal of fuel chemical exergy. In succeeding chapters on fully engineered single fuel cell types, without circulators, irreversibilities in both reformer and cell are discussed. The as yet unattained achievement of good economics involves fuel infrastructure and logistics, power density and cell endurance, and equipment volume and mass, with different compromises for transport, stationary power and the numerous niche markets. Stack arrangements and balance of plant simplicity are vital economic features. The introduction of circulators to these empirically developed systems will represent major engineering.

3.6   IRREVERSIBILITY IN CALCULATION ROUTES

The equilibrium routes which essentially underlie any fuel chemical exergy calculation involve careful avoidance of all irreversibility, so that the routes can be rather unreal (e.g. Figures A.1, A.2 and A.4). These figures however, bring out the major point that equilibrium concentrations must be used in equilibrium diagrams. Any failure to accept that point results in omission of consideration of circulators, or erroneous sizing thereof. See, for example, Kotas (1995), Figure 2.14, p. 46.

In Figure A.4 the fuels are consumed in separate CO and H_2 cells to avoid mixed potential irreversibility. The alternative irreversible mixed process, using one fuel cell, involves:

1. A mixing loss as the fuels are mixed. Separative work would be required to undo the mixing.

2. A mixed electrochemical reaction with a mixed potential difference, inevitably of reduced output. Detailed consideration of the breakdown of the oxygen reduction and hydrogen oxidation paths would be involved in a calculation, including any chemically irreversible cul-de-sacs, such as platinum particles insulated from the electrode, which catalyse irreversible combustion rather than reversible electrochemical oxidation.

3. Mixed expansion to atmospheric conditions, which could not simultaneously achieve the correct discharge conditions for each of the two products, steam and carbon dioxide.

Calculation of the efficiency of the practical air-breathing fuel cell, based on the fuel calorific value, involves ignoring or not realising that combustion is irreversible, with destruction of the kinetic exergy of high-speed product molecules.

3.7 JUGGLING WITH IRREVERSIBILITIES

At a more sophisticated level, it may be possible to use one loss mechanism to generate an effect which greatly reduces another loss. In the Japanese experimental PEFC of Masahiro *et al* (1998), hydrogen diffusing through Nafion is oxidised (non-Faradaically) to water, with the aid of platinum particles distributed in the Nafion. The water is stored in titanium dioxide particles, again within the Nafion, and used to forestall drying, and so to enhance and steady the ionic conductivity of the Nafion. The result is a large increase in available current.

Appendix A shows that for equilibrium fuel cells, high-temperature operation offers no advantage, but indeed a theoretical disadvantage. However, in a practical non-equilibrium set-up, where unused fuel and unconsumed oxygen are features and these have to be combusted in a gas turbine, the high-temperature fuel cell is likely to be at an advantage. Further discussion appears on this point in Chapters 4 and 5, in respect of the SOFC and the MCFC. The PEFC (Chapter 6) must, for lack of high-temperature heat, have combustion-fired reformers carrying the thermodynamic penalty of combustion irreversibility. An alternative way of making this assertion is that a theoretical electrically driven reformer must be used in the low-temperature calculation route of Figure A.4.

MCFCs and intermediate-temperature SOFCs can incorporate catalysed reform at their anodes, where the hydrogen electrochemical oxidation proceeds simultaneously, and heats the non-Faradaic and endothermic reform and shift reactions! The latter process is immediately superior to a separate reformer, because it eliminates combustion reaction irreversibility. Heat produced at such an anode is given, in Appendix A, the title 'reversible heat', that is heat produced without the thermal degradation which occurs in the combustion reaction.

At 600 °C in the MCFC, the dynamic equilibrium conditions are ideal for anode reform. The voracious oxidation reaction swallows both reform and shift reaction products as they are formed. The latter reactions are left striving to equilibrate. In the high-temperature SOFC the reform reaction is very vigorous, and uneven temperature distribution can occur. To avoid that irreversibility, Siemens Westinghouse still employs separate reformers. More irreversibility, but SOFC temperatures are on their way down! The intermediate-temperature SOFC is emerging.

Whereas the hot systems can consume CO, the cool systems suffer CO-poisoned platinum catalysts, and must have a shift reactor to consume the CO. Platinum poisoning is an irreversibility. The alkaline fuel cell (AFC), although without platinum, is especially incompatible with CO because of its KOH electrolyte. It needs a pure hydrogen fuel, and air with CO removed. The latter two purifications carry their own irreversibilities.

3.8 AIR-BREATHING FUEL CELLS – IRREVERSIBILITIES

In Appendix A, calculations show a status, for fuel cell isothermal Faradaic oxidation, of a high vacuum of reactants relative to a high concentration of product. That calculated status cannot even be approached in the laboratory, for lack of adequate semi-permeable membranes and circulators (concentration cells). The equilibrium fuel cell of Figure A.1 is dead-ended, whereas the air-breathing open-ended design must have both of its electrodes swept by a parallel flow, with an inlet and an exhaust. The decision to use air breathing eliminates the circulators of Figures A.1 and A.2, precludes the achievement of equilibrium concentrations of reactants, and imposes irreversible diffusion in fuel and oxygen concentration gradients. Economic decisions have to be made to select the stoichiometry, the fraction of unused cathode exhaust oxygen and the fraction of unused anode exhaust fuel. The fraction of unused fuel, which must be irreversibly combusted, is a major variable. In low-temperature systems requiring a reformer, fuel from the cell exhaust is combusted to heat the reformer. The solid oxide system can consume natural gas, at its hot anode, by combining reform and isothermal oxidation. The hot, fuel-rich exhaust leads immediately to the notion of integration with a gas turbine, involving pressurised fuel cell operation. The objective is best utilisation of the electrochemically unusable exhaust fuel from the anode of the fuel cell, and its hot depleted air cathode exhaust. In a future very large plant, a combined gas turbine/steam turbine cycle could be used, or the gas turbine exhaust could be coupled to a thermal process. Since the engine is using the calorific value of surplus fuel, to produce marginal power, it is effectively a scavenger in the backyard of the fuel cell. Whereas the fuel cell gets the opportunity to use much of the chemical exergy of the fuel, the Carnot limited engine can only use part of the calorific value of its leftover fuel. That is the sensible thing to do, but it would have been much better if the fuel utilisation of the fuel cell could have been improved to the extent of making the gas turbine uneconomic. Better fuel utilisation by the fuel cell is not a widely discussed objective, perhaps because of confusion leading to erroneous priorities, arising from efficiency assessments on a first-law, calorific value basis, rather than a second-law basis, which close the gap between fuel cell and engine efficiency.

As demonstrated in Appendix A, the high-temperature solid oxide fuel cell does not, at equilibrium, have a thermodynamic advantage, neither in its equilibrium form (Figure A.1), nor in its practical air-breathing form. The decline of the fuel cell electrical power output (with increased temperature) is balanced by an equal increase in Carnot cycle power output. However, there is a practical advantage in going to high temperature in that overvoltages are reduced, and efficiency rises, via smaller irreversibility. Moreover, the need for catalysis declines and expensive platinum is not required to achieve high power density. The temperature is high because of the need for high solid state electrical conductivity and minimum ohmic power loss. In practical IT/SOFC arrangements, electrode reform of natural gas and direct utilisation of

CO and H_2 is of great advantage, shared with the MCFC. These factors will recur in the separate chapters on each cell type.

In the MCFC there must be carbon dioxide recirculation from anode to cathode, as per Figure 5.3. The recirculation of anode gas as a means of getting at the carbon dioxide represents an irreversibility, since additional circulator power must be committed, and additional pressure drop losses encountered.

3.9 LIQUID ELECTROLYTES AT THE ELECTRODE, 'ICE' FILMS, MARANGONI FORCES AND DIFFUSION IRREVERSIBILITIES

The three-phase interface is a well-known concept, notably in connection with the invention by Bacon of bi-porous electrodes (Bacon, 1969). At equilibrium, in such electrodes, gas, liquid electrolyte and catalytic solid are in contact at a convoluted meniscus at the coarse-pore, liquid-containing region interface with the fine-pore, gas-containing region. The picture changes with departure from equilibrium. The meniscus becomes mobile under the influence of surface tension gradients or Marangoni forces – an invisible complex.

In Grahame (1957), the presence at a liquid surface of a quasi-ice structure was hypothesised. Moreover, study of droplets on inclined planes, and of bubbles clinging to vertical surfaces, reveals behaviour (contact angle hysteresis) which cannot be accounted for without such an ice film. On a vertical wall, a bubble experiences an upward buoyancy force, but in the absence of an ice film, a zero restraining force. The restraining force is made evident by the phenomenon of contact angle hysteresis, followed by film rupture as bubble growth occurs, and leads to bubble detachment.

As soon as equilibrium is abandoned and an electric current flows, say in a PAFC, reactants and products at the electrically connected multifaceted catalyst interface give rise to concentration gradients, which are the driving force for diffusion and the source of Marangoni forces (Guelcher et al., 1998). Once the surface tension becomes dynamic, equilibrium concepts such as contact angle lose their validity. The latter is demonstrated by the well-known 'tears of strong wine'. A glass of neat spirits, whisky or brandy, left to settle, forms a continuously moving supermeniscus much taller than a water meniscus. Low-alcohol water piles up at the apex of the supermeniscus with an intriguing cellular pattern of down-flow legs and leg bouncing at the base. The whole irreversible phenomenon is derived from alcohol evaporation and driven by the surface tension gradient. The interfacial shear stresses are obviously sufficient to rupture the quasi-ice film. For lack of data to analyse, little more can be said than the preceding qualitative description. The existence of multifaceted catalyst crystallites, with varied catalytic effectiveness, must cause surface tension gradients and shear-stress-driven currents. Even an inert non-wetting substance like PTFE will be affected near the interface by concentration gradients. Unfortunately the electrolyte behaviour within a bi-porous electrode is not observable. Guelcher *et al.*'s (1998) paper is a modern account of surface tension gradient effects in electrochemistry. Such effects could be relevant to demonstrated, but unexplained, relationships between catalyst particle size and catalyst performance (Savadogo and Essalik, 1996).

The theory of the mercury-drop electrode and the potential for zero charge (pzc) is affected by the existence of a film of quasi-ice at the mercury/aqueous electrolyte interface. More than mere surface tension is involved, but little can be made numerate in the absence of properties for the film.

Surface tension gradient effects add to the better known phenomena of density-gradient-driven convection, concentration-gradient-driven diffusion and electrical-potential-gradient-driven ion migration, which appear in the existing theory of cells and electrodes. The potential difference of a working cell is affected by all the near electrode effects mentioned here. The experimental and analytical difficulty is to separate the variables. Indeed the fluid mechanical effects stir the electrochemical reaction, and make cause and effect difficult to discern.

3.10 OVERVOLTAGE – AN ELECTRICAL IRREVERSIBILITY

At the three-phase interface of an aqueous fuel cell, electrons must transfer from the electrically connected catalyst to, for example, hydronium ions (H_3O^+) within a hydration sphere in the electrolyte at the solid/liquid interface, where there is a steep electrical potential gradient. The ion and its impedimenta, located in the potential gradient, are in thermal agitation or vibration. At a favourable configuration of the vibration, electron transfer can occur through the electrolyte, giving rise at equilibrium to a balanced or isotropic exchange current. Away from equilibrium, with an unbalanced electrode current, the electrochemical potential difference, electrode to electrolyte, drives the current. The chemical potentials due to concentration differences act together with electrical potential differences to become the electro-chemical potential difference. Work is done on the electrons to achieve current generation, via

$$2H_3O^+ + O_2 + 2e^- = 3H_2O$$

The achievement of unidirectional current consumes some work, increasing with the current, an irreversibility called overvoltage. The historic plating bath experience was that plating action awaited the imposition of a voltage significantly higher than the open-circuit volt-age. For minimisation of overvoltage, the part played by electrocataly-sis is experimentally evident. Exchange currents can be maximised by choice of catalyst material or configuration. Catalyst theory, involving activated complexes (Atkins, 1995), is not completely satisfying, since it does not, for example, account for the beneficial effects of disloca-tions in platinum crystals. Electrocatalysis is known to be promoted by hydrogen atoms adsorbed on the platinum surface. (Catalytic action ceases when the hydrogen atoms are replaced by CO.) The part played by tunnelling electrons in the electrolyte is obscure, but they clearly influ-ence the electrochemical potential gradient in the double layer, where the electrochemical action is concentrated.

The widely used Tafel plot (Koryta, 1993) of log current density against overpotential yields information about the magnitude of the exchange current.

For high-temperature fuel cells without aqueous effects, overvoltages are still present but at reduced magnitudes. For a single fuel with a cal-culable unique V_n (e.g. 1.23V for standard conditions hydrogen), there remains the fact (see the introduction) that the equilibrium electrode concentrations of fuel and oxygen are very low, and those must readjust substantially upon moving away from equilibrium, without circulators. The terminal voltage must fall! That effect is also present for aque-ous cells, too. Mixed fuels imply non-equilibrium mixed potentials and circulating currents dependent on electrode surface details.

3.11   BICONDUCTOR LAYERS AT THE ELECTRODE/ELECTROLYTE INTERFACE

The three-phase interface can be favourably influenced by the introduc-tion of a thin layer of material, capable of conducting both ions and electrons, between the electrolyte and the electrode structure.

For the SOFC, for example, the use of uranium oxide has been studied but not adopted. The onset of current cut-off by concentration polarisation can be delayed, so that the maximum current obtainable is increased, a significant reduction of irreversibility. SOFC cathodes now almost universally use a coating of lithium strontium manganate, a joint electron/ion conductor.

For the PEFC, low-platinum loading electrodes are being developed (Kumar *et al.*, 1995), which, for optimum electrocatalysis, seek to achieve a maximum fraction of platinum crystallites which are electrically connected to the external circuit. In the PEFC, Masahiro, *et al.* (1998) clearly demonstrate that platinum crystallites which are not electrically connected to the carbon/graphite promote non-Faradaic oxidation of hydrogen molecules diffusing in the electrolyte, an unhelpful irreversibility if it occurs in the same region as Faradaic oxidation, and water production. In Foster (1998), the proposal is investigated to increase the fraction of electrically connected crystallites using electrolyte conductor 'wires', in the interface region.

3.12 IR DROP

The migration of ions in the electrical potential gradient is another work-consuming, heat-generating process. The potential gradient is revealed (Figure 2 of Ralph, 1997) by open-circuiting a working galvanic cell. The gradient approximates to linear and is called an IR drop. Another irreversibility has been introduced.

3.13 REMARKS

Any engineered system achieves economic performance by accepting losses which are too expensive to eliminate, and by minimising, rather than eliminating, other losses. The subject must involve knowledge of loss mechanisms or irreversibilities, via exergy theory. The world fuel cell industry faces a challenge, as mentioned in the foreword, which cannot be ignored.

4

Solid Oxide Fuel Cells (SOFCs)

Happy he, who has availed to read the causes of things.

Virgil, 70–19 BC

In a qualification of the above quotation, this book points out that 'the causes of things' in the fuel cell industry literature are written without a logical thermodynamic basis. The author introduces such a basis, namely reversible isothermal oxidation.

In this chapter, the Harvard-style references, many of which are patents, are grouped by manufacturer, and also appear in the main list of references at the end of the book.

4.1 INTRODUCTION

4.1.1 The SOFC

The SOFCs described in this chapter are 'incomplete' systems. Some SOFCs which use hydrogen are doubly 'incomplete' in the absence of a 'hydrogen mine' or source of low-cost hydrogen. Remarkably, the development of the hydrogen mine gets no mention in the programme of the 2005 Grove Symposium in London. The *Journal of Power Sources*, containing the Seventh Grove Symposium in Vol. 151,

Fuel Cells, Engines and Hydrogen – An Exergy Approach Frederick J. Barclay
© 2006 John Wiley & Sons, Ltd

October 2005, also fails to mention the problem. The industry still does not seem to realise the vitally important crossroads at which it is located, and the cause of its huge losses, as highlighted in the Price Waterhouse review of the fuel cell industry (Price Waterhouse Coopers, 2005).

The SOFC system, however, has the potential to use natural gas directly, and thereby the opportunity to bypass the hydrogen source problem for stationary non-vehicle applications. The North Western University US patents on direct hydrocarbon oxidation are 2001/6,214,485 B1, 2002/6,479,178 B2, 2003/0,118,879 and 2004/0,033,405, which deal with special catalysts and anodes. The parallel work at the University of Pennsylvania, http://www.upenn.edu/, is given in McIntosh S and Gorte (2003), correspondence address gorte@seas.upenn.edu/.

A vehicle **must** carry stored hydrogen, and for one version of the proposed Renault fuel cell a compact liquid hydrogen tank has been developed by Air Liquide.

The beginnings of the SOFC are recorded in an early East German University patent (Möbius and Roland, 1968) which shows awareness of many of the variables still being worked upon today. The oxides of lanthanum, zirconium, yttrium, samarium, europium, terbium, ytterbium, cerium and calcium are mentioned as candidate electrolyte materials. The proposed monolithic planar arrangement has, however, been abandoned by many companies, on the example of Allied Signal. One notable exception is a reversion to a circular planar concept by Ceramic Fuel Cells of Australia, UK (Section 4.7). The Rolls-Royce all-ceramic fuel cell (Section 4.3), which is monolithic and has one compliant feature, namely a gap, is a major exception. One modern trend is towards lower SOFC temperatures, with the intermediate-temperature IT/SOFC allowing the use of cell and stack arrangements with some flexibility and manoeuvrability based on new electrolytes, metallic flow plates, electrodes and interconnects.

The small-tube SOFC (Section 4.9) remains outside the latter trend, since it is already very manoeuvrable and indeed shockable.

The UK Ceres Power fuel cell based on ferritic stainless tubes or plates and a low-temperature electrolyte may also approach being shockable. (See Section 4.6 and the listed company web site. Also see the company history by Bance *et al.*, 2004.)

Siemens Westinghouse is a major player, with its all-ceramic large-diameter tubular fuel cells. These are to be deployed with integrated gas turbines (Section 4.2).

4.1.2 Electrolytes

Solid electrolytes such as yttria-stabilised zirconia (YSZ) have emerged from a historic sorting process. The Russian geologist Count L. A. Perovsky (*Encyclopaedia Britannica*) gave his name to the mineral perovskite ($CaTiO_3$). The mineral $MgSiO_3$ is also called perovskite, and is an abundant constituent of the earth's crust. These minerals, of generic and stoichiometric formula ABO_3, share a common orthorhombic structure, which may be reproduced when alkali metals or rare earths (Möbius and Roland, 1968) are substituted, during electrolyte development, at the A and/or the B position. For example, the perovskite material $BaTiO_3$ is used in electronic engineering. Below, Grimble (1988) and Zymboly (1988) give accounts of two new intermediate-temperature, non-stoichiometric, perovskite electrolytes for IT/SOFCs.

4.1.3 Electrolyte Thickness

In the SOFC, the thickness of the electrolyte layer is a vital consideration, as is any additional biconductor interface; 20–50 μm is referred to as a thick electrolyte film, a design compromise between the minimisation of electrical losses and freedom from defects. Thin films may go down to nanometres and are more difficult to make without defects. Global Thermoelectric (Kuo *et al.*, 1999) is in SOFC production at 6 μm and InDEC (Section 4.12) at 5 μm. In the small-tube SOFC the membrane thickness is about half a millimetre. It is difficult to see the latter thickness coming down for improved efficiency to 6 μm, impossibly delicate tubing.

Matching the coefficients of expansion of electrode structure, electrolyte and interconnect is a development problem for all SOFC designers pursuing good resistance to thermal cycling or thermal shock.

4.1.4 Cell Performance

The internal performance of a contemporary air-breathing fuel cell, without a potential difference across circulators/concentration cells, is summarised by its voltage–current characteristic, see Figure 4.6.

With its reactant and product connections isolated, but equilibrium concentrations remaining in the cell, the cell voltage is a maximum, namely V_n, at open-circuit, zero-flow equilibrium (Table A.1). Equilibrium entails the presence of a thermodynamically reversible, symmetrical or balanced exchange current, described by Marcus (1964; 1982) as being an equilibrium exchange between electrode electrons and vibrating hydrated ions in the thermally agitated electrolyte. Electron exchange alternates to and fro depending on the phase angle of the hydrated ion oscillation. When reactants and product are allowed to flow, and load current is drawn, the exchange current is unbalanced then overwhelmed. In Figure 5.6, in the absence of circulators, there are, upon moving away from equilibrium with its very low fuel and oxygen concentrations at the electrodes, large concentration changes upon entering the diffusion regime which force the cell voltage to drop steeply and non-linearly. The latter is the author's unique explanation of the steep drop, which has not previously been given. (See also Section 3.4.) The steep drop is termed overvoltage, a historic term from plating experience, in which plating did not commence until a voltage significantly in excess of open-circuit equilibrium was applied. There follows an ion conduction region with a near linear, small-slope. 'IR voltage drop', until concentration polarisation is encountered, and the voltage drops steeply to zero. An advantage of SOFCs is that with high-temperature operation, overvoltage magnitudes are down relative to other systems. In the languare of Chapter 3, irreversibility is reduced!

See Atkins (1995), Koryta (1991; 1993) and Bard and Falkener (1980) for the Butler–Volmer theory and Tafel slopes.

4.1.5 Competitive Cells

A competitive cell is likely to have in its V/I curve (Figure 6.5) a small IR drop (thin electrolyte film) and an extended current range before concentration polarisation sets in. In the SOFC, moreover, an expensive platinum catalyst is avoided due to high reaction rates at high temperature. Engineering for mass production is very important to the achievement of low capital cost, for example the Rolls-Royce fuel cell (Section 4.3). Its compact stack arrangement is described in US patent 2003/0,096,147 A1, an improved version of the previous patents.

4.1.6 Oxygen Ion Concentration

The SOFC has been wrongly described in the early literature as an oxygen concentration cell. On the contrary, the author asserts that all fuel cells are ion concentration devices. Also that zero external current cell equilibrium equates to an internal balanced state, at which no net ion transfer occurs between electrodes, due to a balance between counter-directed ion migration and diffusion. The SOFC (Möbius and

Roland, 1968) is special because it uses O^{2-} ions, mobile in the solid electrolyte and formed at the cathode, within the overall reaction

$$O_2 + 4e^- = 2O^{2-}$$

Another doubtful historic assertion (Min, 1995) is that the cathode reaction reverses at the anode, followed by

$$O_2 + 2H_2 = 2H_2O$$

However, the latter reaction is irreversible combustion! It is undoubtedly correct to stick with Faradaic work producing charge transfer oxidation at the anode:

$$O^{2-} + 2H = H_2O + 2e^-$$

The SOFC is then an oxygen ion concentration device.

4.1.7 Unused Fuel

The inevitable unused fuel and depleted air in the practical fuel cell hot discharge have, for economy, to be exploited in the combustion chamber of a steam generator (small scale and present), or of an integrated gas turbine (large scale and future) (Domeracki *et al.*, 1995; George *et al.*, 2001). The combustion of unused fuel and utilisation of the resulting heat in an engine must be seen as Carnot limited, and relatively inefficient, compared with isothermal electrochemical oxidation in the fuel cell, the senior partner of the integrated, or hybrid, power generation process. The latter picture is not visible in early work in which the fuel cell produces a minority of the power. This is very far from the ideal: the Carnot limited engine using surplus fuel produces more power than the fuel cell which discards the surplus fuel. Conceptually, the gas turbine is a scavenger, feeding on the exhaust from a primary power producing fuel cell.

The gas turbine compressor is the means of heating the SOFC to ion conduction operating temperature, a considerable advantage, amplified in Section 4.1.9.

4.1.8 SOFC Internal Process

Repeating from earlier, an equilibrium fuel cell directly generates an electrical potential difference, V_n, proportional to the work function, the

Gibbs potential, the units of which are related to the electrical potential by Nernst's equation:

$$V_n = -\Delta G/(\nu F), \quad \Delta G = -RT\ln K, \quad V_n = RT\ln K/(\nu F) \qquad (4.1)$$

There are losses (energy production from exergy annihilation), but no energy conversion. Carnot limited energy conversion does not occur in a fuel cell, and is exclusive to the integrated gas turbine! The confusion arises from not defining energy precisely, energy being the sum of the chaotically interactive translation, rotation and vibration of molecules **in a solid or fluid**. Kinetic energy is a misnomer for kinetic exergy; likewise, potential energy is a misnomer for potential exergy. The latter are totally available, and not Carnot limited.

4.1.9 SOFC Preheating for Start-Up

It is a characteristic of the SOFC that, in contrast to the PEFC, it must be substantially preheated. When the cell stack is integrated with a gas turbine, initial containment pressurisation and preheating to compressor outlet temperature can conveniently be achieved by spinning the gas turbine compressor, using grid power. Fuel can then be injected into the gas turbine combustion chamber, to produce heated combustion chambers and a hot exhaust. The fuel cells will start when their electrolyte is hot enough to be ionically conductive. The start-up procedures will differ over the range of SOFCs (and MCFCs) in existence. Small stacks without an integrated gas turbine have alternative arrangements, using start-up electric heaters. (For example, see Section 4.9, on the Global Thermoelectric module.) Following on immediately in Figure 4.1 is a modern proposal for gas turbine integration from Siemens Westinghouse.

4.1.10 SOFC Manoeuvrability

For all-ceramic HT/SOFCs the rate of preheating will be slow compared with IT/SOFCs with flexible metallic structural parts between the MEAs. In future SOFCs capable of direct oxidation of hydrocarbons (Section 4.1.11) and without a reformer, the preheating manoeuvres will be simplified by the absence of the reformer. An exception is the small-tube SOFC (Sections 4.11 and 4.12), which can be preheated rapidly.

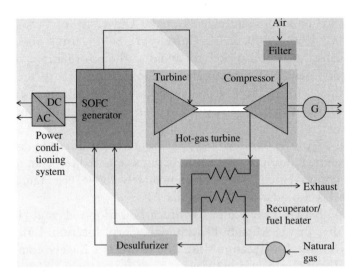

Figure 4.1 Siemens Westinghouse SOFC/gas turbine integration scheme

4.1.11 Direct Hydrocarbon Oxidation

Perry-Murray *et al.* (1999), Park *et al.* (1999; 2000; 2001) and Gorte *et al.* (2000) cover the direct oxidation of hydrocarbons such as methane in IT/SOFCs. This is a new development which could destabilise the existing competitive equilibrium. The reform step is eliminated! The thermodynamics of equilibrium electrochemical oxidation of methane, and hence of natural gas, are favourable (Chase *et al.*, 1985; 1998), there being minimal reversible entropy growth. That could lead to production of electricity with little heat, and without a large natural gas reformer, the latter being a major simplification, and economic gains. The minimum heat ideal then recedes, because gas turbine integration remains necessary to scavenge and burn unused fuel from the fuel cell exhaust.

In a 1999 letter to *Nature* (Perry Murray *et al.*, 1999), from North Western University, Illinois, the authors record the first laboratory achievement of useful oxidation rates for direct methane electrochemical oxidation, using an IT/SOFC. The cathode structures were porous lanthanum strontium manganite (LSM) on porous $(Y_2O_3)_{0.15}(CeO_2)_{0.85}$ or YDC. The anodes were cermets, porous YSZ with nickel in the pores. The laboratory operating temperatures were in the range 500–700 °C. The account of the North Western work, reporting on new anode types, continues on pp. 921–924 of Williams (2002).

There are parallel achievements at the University of Pennsylvania, (Park *et al.*, 1999; 2000; 2001; Gorte *et al.*, 2000), using anodes with copper substituted for nickel to avoid carbon formation. The last two papers include the electrochemical oxidation of dry fuels other than methane, for example gasoline and diesel, the chemical exergy of which is difficult to calculate, since they are mixtures requiring separative work.

At p. 307 of Williams (2002), MIT enters the direct hydrocarbon field presenting alternative anode structures, asserted to be an improvement on copper-based anodes, as immediately above. See Figure A.5 on methane direct oxidation in Appendix A.

A fusion between these direct hydrocarbon proposals and IT/SOFCs such as that of the Mitsubishi Materials Corp. (Section 4.5), or that of Ceres Power Ltd (Section 4.6), could make a hugely competitive, simplified and cheaper future system, with further major development potential to take account of the points in Appendix A of this book, that is fuel cells allied to concentration cell circulators, namely complete fuel cells. Such a development will await judgement of this book, and departure into the history of the application of combustion theory to fuel cells using isothermal oxidation.

Imperial College's Centre for Ion Conducting Membranes is working on perm-selective membranes, and perhaps could develop the concentration cells mentioned in Appendix A (Williams, 2002), pp. 878–891.

4.2 SIEMENS WESTINGHOUSE

The reader should note that the tubular SOFC at 1000°C is too hot (excessive reaction rates) for successful reform at the anode, so that separate reformers must be employed. The MCFC of Chapter 5 at 600°C can use anode reform, without high anode thermal stress, such as would occur in the SOFC of Figure 4.2.

A consistent, steady, approach by Siemens Westinghouse has been made, based on the arrangement shown in Figure 4.2, from the patent (Baozhen Li *et al.*, 2001). The operating temperature has stayed at 1000°C, the optimum for the electrolyte, which has remained as YSZ.

In a possible future development (Singhal, 2000), pp. 305–313, aimed at high power density, the circular tube is flattened to become a rectangle with rounded corners. The flat sides are supported by internal ribs, which strengthen the rectangle against stacking of one cell above another, and provide useful longitudinal passages. The entire topside of

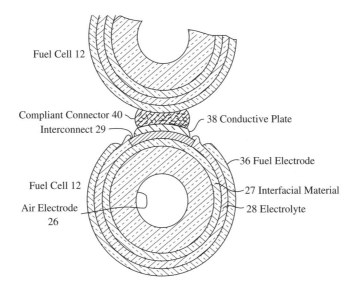

Fuel Cell 12

Compliant Connector 40
Interconnect 29

38 Conductive Plate

36 Fuel Electrode

Fuel Cell 12

27 Interfacial Material

Air Electrode
26

28 Electrolyte

Figure 4.2   High-temperature, tubular, all-ceramic SOFC

the rectangle is covered by an LSM interconnect of large area, minimal thickness and electrical resistance. This concept has not appeared in a real installation.

In Baozhen Li *et al.* (2001), however, Siemens Westinghouse enter, under competitive pressure, the ITSOFC field, at 1000–750 °C, with fresh materials science: namely, biconductor films based on terbium or scandium, between the cathode and electrolyte, resulting in a thicker boundary (TPBR) and increased current density. At the interface between the biconductor and the electrolyte there will still be a thin electrolyte region containing tunnelling electrons. The fuel cell oxygen reduction cathode is generally recognised as being in greatest need of improvement, and not only in SOFCs. There is no sign, however, of such cells (Baozhen-Li *et al.*, 2001) appearing in Siemens Westinghouse gas turbine/SOFC hybrids.

4.2.1   Siemens – SOFC Integration with Gas Turbines

Siemens Westinghouse is advanced in its pursuit of gas turbine integration. High-temperature depleted air and unused fuel from the fuel cell cannot be discarded, and are best used in a gas turbine, as outlined in Baozhen Li *et al.* (2001), the Siemens patent for a Single Module Pressurised Fuel Cell Turbine Generator System (pp. 974–979). The effect of pressure on the efficiency of integrated fuel cell gas turbine arrangements is not entirely clear. The fugacity-based JANAF tables (Chase *et al.*, 1985), as displayed in the introduction, show that the equilibrium open-circuit potential V_n reduces with pressure for hydrogen fuel, but is constant for carbon monoxide fuel. Basel and Pierre (1995) show from experimental results with hydrogen on the Westinghouse rig at Toronto Hydro that, at power and away from equilibrium, the fuel cell power improves with pressure. Many parameters are not stated. Irreversibilities, such as the pressure drop in flow plates and overvoltage, are possible variables. The improvement is large between 1 and 3 bar, and trivial between 10 and 15 bar. There must be an unstated gas turbine power reduction, not accounted for in the rig.

Siemens information on its concept for a combined fuel cell/gas turbine plant is available at http://www.powergeneration.siemens.com/en/index.cfm/.

4.3   ROLLS-ROYCE

Adherence to ceramic materials and 1000 °C operating temperature is displayed by Rolls-Royce (see web site) which offers direct competition to Siemens Westinghouse via its integrated planar cell arrangements, with thick (40 μm) electrolyte layers. The arrangements of the cell and stack are described in two overlapping patents (Gardiner *et al.*, 1996; 1997), which, for overall understanding, are best read together. Moreover, the detailed tubing arrangement within the rectangular cell stack, flanked at each end by pre-reformers, needs careful appreciation after repeated reading and examination of the figures in the Rolls-Royce patents. Printing out of the patents and their figures via their US patent numbers and the procedure at the end of Chapter 2 is recommended. However, a recent US patent, 2003/0,096,147 A1, describes an improved stack arrangement, which overtakes the latter advice.

The Rolls-Royce fuel cell modules and stacks are devoid of compliant features, as the cross-section in Figure 4.3 shows, unless one counts the gap as compliant. Cell improvement is clearly possible via thin (say 6 μm) electrolyte layers, as employed in a planar IT/SOFC by Global Thermoelectric (Section 4.9) and Steele *et al.* (2000a, b). Alternatively, modern reduced temperature electrolytes could be used.

There is no sign of such changes from Rolls-Royce, whose module of cells shown in Figure 4.4 has not been changed over the years.

The photograph, by courtesy of Rolls-Royce, shows a double-sided module of 50 W fuel cells corresponding to Figure 4.3. The method of

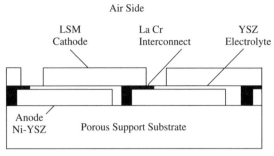

Figure 4.3 Cross-section of Rolls-Royce 50 W fuel cells showing internal structure

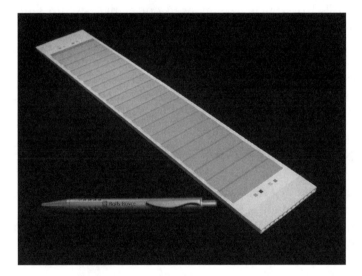

Figure 4.4   Rolls-Royce 50 W fuel cells in twin module of 20 per side

compact arrangement of these modules into integrated planar stacks is obtainable from their patents, which are sometimes difficult to read. There are 20 000 cells per megawatt!

An immediate comparison with the Siemens Westinghouse tubular cell reveals that the Rolls-Royce arrangement is likely to be relatively compact. In Figure 4.4 cells, gas passages and dimensions are apparent.

Envisage a vertical slab of four modules with long axes horizontal. Fuel tubes, with penetrations half-way down, are vertical, on both flanks of the slab, whilst air tubes penetrate horizontally the gap between the two fuel cell layers of the modules. All tube sets are manifolded, and connected to the system. Many such tube-equipped slabs make up four sub-stacks, within a main 2.5 MWe stack of 25 000 cells. Larger installations can be built up of 2.5 MWe units. The main reform process is at the catalyst within the vertical, cell-heated fuel tubes. An ejector provides for depleted fuel gas recirculation.

The main aim of Rolls-Royce, shared with Siemens Westinghouse (George *et al.*, 2001), is fuel cell/gas turbine integration in large-capacity pressurised units. In the longer term, combined electricity and synthesis-gas production can be supported.

A modern synthesis gas plant can be coupled to a gas/steam combined cycle. The gases could in a future installation be consumed at enhanced efficiency by a hybrid fuel cell/combined cycle plant, notably if the fuel cells were equipped with concentration cell circulators. See Section 5.2.13.

The Rolls-Royce plan was mapped out at the Seventh Grove Symposium (Agnew 2001), and is based on optimisation of the current, all-ceramic design as part of a gas turbine hybrid. The Rolls-Royce gas turbine will be optimised for hybrid service. The potential of its fuel cell, for cheapness via mass production, is stressed in its patents. Capital costs could come down to $300/kWe. There are no signs up to 2005 of a Rolls-Royce ITSOFC waiting in the wings.

4.4 NGK INSULATORS

NGK Insulators in Japan had a patent for a system closely resembling that of Siemens Westinghouse. NGK Insulators has sold out to Mitsubishi, and dropped out of the fuel cell business, an interesting piece of history.

4.5 MITSUBISHI MATERIALS CORPORATION (MMTL)

In a press release dated 17 December 2001, see listed web site, MMTL claimed a new design of IT/SOFC having 1.8 W cm^{-2} at 800 °C, a world-record high at the time. Partners are Kansai Electric and Japan Fine Ceramics Centre. The Japanese cells are to be compared with the German FZJ cells described in the press release of Section 4.8, below. A major increase in power density comes from a new electrolyte, namely $La_{0.9}Sr_{0.1}Ga_{0.9}Mg_{0.1}O_{2..9}$.

The MMTL cell (Figure 4.5) is flat, circular and with a 154 mm diameter. These advanced cells have the temperature range 650–800 °C,

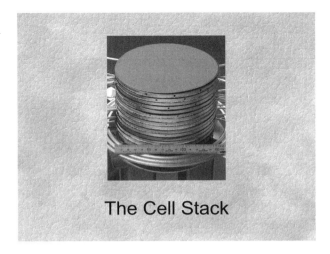

The Cell Stack

Figure 4.5   MMTL cell stack

and, by the author's assertion, are candidates to use the direct methane oxidation technology of Section 4.1.11, notably the copper bearing anode of North Western University, or the MIT alternatives (Williams, 2002). These are beginning to approach the performances of Ballard PEFCs as shown in Figure 6.5.

The account of the development of the MMTL advanced cell is continued in Williams (2002), pp. 917–920. The series interconnected cells in the stack are separated by metallic separators and current collectors, along the lines of those abandoned by CFCL (Section 4.7).

A voltage–current characteristic of an MMTL cell is shown in Figure 4.6. The actual electrode area associated with the figure is $177\,cm^2$.

4.6   IMPERIAL COLLEGE LONDON AND CERES POWER LTD

A modern, non-stoichiometric, fuel-cell-oriented, perovskite material, $Ce_{0.9}Gd_{0.1}O_{1..95}$ (COG), has been developed by Professor Steele at Imperial College London, as recorded in Steele *et al.* (2000a, b). Professor Steele has found the way, via COG, to a porous ferritic stainless steel, tubular or flat square IT/SOFC operating at 500 °C, superb achievement.

The cell is incomplete and without circulators, as indeed are all the cells in this and the following chapters. The tubes support the MEAs.

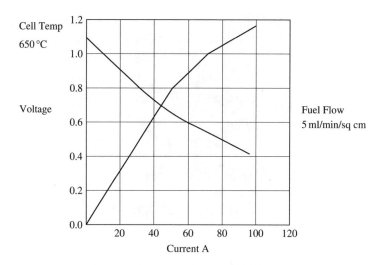

Figure 4.6　MMTL advanced fuel cell *V/I* curve

A company, Ceres Power Limited, see web site, has been set up to exploit the new fuel cell, which is described in its flat square form in Steele *et al.* (2001). In this reference, the efficiency theory is of the type asserted to be incorrect in this book! The initial aim of the Ceres cell is anode reform, aided by a pre-reformer, and, optionally, integration with modern small gas turbines. A replacement for the car battery is an immediate target. Direct hydrocarbon oxidation is over the horizon.

4.7　CERAMIC FUEL CELLS LTD, AUSTRALIA

In Figure 4.7, 800 °C is the operating temperature of the planar IT/SOFC, patented by Ceramic Fuel Cells Ltd of Australia (Gibson *et al.*, 1995; Wolfe *et al.*, 1999; Badwal *et al.*, 1996; 1997; Donelson *et al.*, 1998).

A special stainless steel was developed in Australia and patented (Jaffray, 1999). Production costs and endurance of the resulting flat plate design (Foger *et al.*, 2000) were improved relative to that of the firm's initial all-ceramic design. The bipolar plates and interconnects were stainless steel. The cell had operational difficulties, and has been discontinued. The firm has just established a UK division, to compete in Europe.

CFCL UK has reverted to an all-ceramic scheme, and is now concentrating its initial production and marketing efforts upon a flat plate

Figure 4.7 CFCL UK fuel cell to compete in Europe

circular fuel cell, arranged in a 40 kW fuel cell stack. CFCL uses a 6×4 array of tall cylindrical stacks with an accompanying balance of plant module. Variables such as electrolyte thickness are not mentioned, and a voltage–current characteristic is not given. However, at the Palm Springs Seminar (Williams, 2002, pp. 964–967) further details of the 40 kW, 850 °C, natural-gas-fuelled first CFCL product were given. The new System Integration Prototype IT/SOFC is of flat, circular, monolithic ceramic construction, as shown in Figure 4.7.

The web site of CFCL UK shows the above construction. The cylindrical cells are 10YSZ electrolyte supported, with LSM cathodes and nickel/10YSZ anodes. Thermal cycling performance is asserted to be good. Tested endurance with zero deterioration is 1000 h.

4.8 FORSCHUNGS ZENTRUM JULICH (FZJ)

The Julich research centre is multiproject and multidisciplinary. It contains, *inter alia*, The Institute for Materials and Processes in Energy Systems (IWV-3). A current illustrated announcement on its web site claims as a historic achievement a 9.2 kW, 800 °C, ferritic steel planar fuel cell with an operating point at 0.755 V at 0.83 A cm^{-2}, or 0.63 W cm^{-2}, using YSZ electrolyte. But see the MMTL claim above for cells using a new electrolyte, $La_{0.9}Sr_{0.2}Ga_{0.8}Mg_{0.15}Co_{0.05}O_{2.8}$ (LSGMC). The latter gives a greatly improved power density of 1.8 W cm^{-2}. See Akikusa *et al.* (2001).

An obtrusive feature of the Julich fuel cell is that its performance on pre-reformed methane is significantly worse than on hydrogen, namely 0.755 V per cell at 0.50 A cm^{-2}. The air stoichiometry is 3.5 as opposed to 2. The fuel utilisation is a mere 59%. None of the promise implied by the circulator-dependent chemical exergies of Chapter 1 is realised in the 'incomplete' cell. The reformer ahead of the German fuel cell provides an output of steam with diluted carbon monoxide and hydrogen as fuel. To consume such a dilute fuel a generous excess of air (adequate oxygen) was needed. The resultant losses, in terms of reduced fuel utilisation, and surplus hot air were overweening. The losses in the fired reformer are not accounted for. Equally the losses involved in hydrogen manufacture are not dealt with, so the detail of the comparison is inadequate. All irreversibilities must be dealt with in the account, see Chapter 3. In Figure A.1 all irreversibilities are eliminated, and the initial appreciation of the fuel cell problem sharpened!

4.9   GLOBAL THERMOELECTRIC

Competition, on similar lines, arises from Global Thermoelectric, Calgary, Canada (see web site), licensee of Forschungs Zentrum Julich (FZJ). Global Thermoelectric produces flat square cell SOFC stacks, in which the MEAs have 6 μm electrolyte. The stacks are integrated with a concentric module, which incorporates preheat of natural gas, water/steam and air, reform, exhaust combustion and heat utilisation, in the twin preheaters. These cells are widely used by other manufacturers for their stacks.

4.10   ALLIED SIGNAL

Whereas most patents refer, irrationally, to the fuel cell as an energy converter (a heat engine is an energy converter), Allied Signal correctly referred to the direct electrochemical generation of electricity (power). The hallmark of the firm, now owned again by Honeywell, was maximum current per unit MEA area.

Allied Signal had a gas turbine integration scheme, in which the gas turbine coped with peak demands without disturbing the output of the fuel cells, and avoiding thermal stress. The intentions for the future of Honeywell are unclear.

4.11   ACUMENTRICS

All the SOFC competitors seek to compete by overcoming the operational limitations of the Siemens Westinghouse SOFC, and a new competitor, Acumentrics Inc., has appeared, to exploit the small-tube SOFC, introduced in Figure 4.8. (See web site list.)

Small YSZ tubes with inked-on internal $La_{0.7}Sr_{0.3}MnO_3$ (LSM) cathodes and nickel zirconia cermet anodes are largely immune to thermal shock, which (above) is a development problem of the Siemens Westinghouse and Rolls-Royce SOFCs. The Acumentrics small-tube system has started commercialisation as the power source of an uninterruptible power system (UPS) of 2 kW (see web site). The natural-gas-fuelled UPS was supplied by Acumentrics to a commercial partner. The web site refers to a future 'super cell', a 100 kW UPS, allied to a flywheel. From the same source 250 kW appears to be the largest size contemplated. The Acumentrics web site is developing rapidly. It emphasises that the small-tube fuel cell stacks are fabricated in-house. A top power of 100 kW is stated, with simultaneous heat production, and with undetectable levels of SO_x and NO_x. The mechanical arrangements adopted to construct a free thermal expansion stack of many cell tubes with preheater and reformer are described in diagrams on the listed web site. Note that the vigorous reform reaction at 1000 °C, which generates unacceptable thermal stress in the Siemens Westinghouse fuel cell, is not a problem for the small-tube SOFC. The thick electrolyte will limit the range of applications. The patent (Kendall *et al.*, 2002) is a link between Acumentrics and Adelan (see next section).

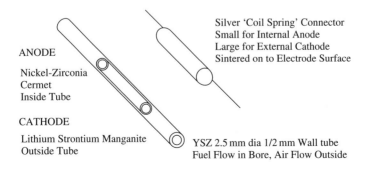

Figure 4.8   Small-tube fuel cell

4.12   ADELAN

Adelan, based at the University of Birmingham under Professor Kendall, is a small UK company, older than Acumentrics and a protagonist of the small-tube SOFC. Adelan is linked historically to Acumentrics, because Acumentrics acquired (Kendall *et al.*, 2002) an Adelan patent. Adelan has subsequent independent patents pending. It has an initial objective of providing fuel cell power and heat for large trucks, and has looked at the development of technology to enable fuel cell and engine to use the same diesel fuel.

4.13   SULZER HEXIS

Sulzer Hexis went into bankruptcy in late 2005, so the following account is history. The Sulzer Hexis fuel cell was exclusively aimed at the domestic market (Batawi 1996; 1999; 2001; Schuler, 2001; Ballhausen, 2001). After a long development history the project matured as the Sulzer Hexis Premiere, an incomplete fuel cell, calculated using calorific value theory.

From the historic patents it is clear that advanced materials science was involved, and that there had been a move into intermediate-temperature operation, to cheapen the cells. For a cell aimed at domestic electricity and heat supply, lowering the cell temperature tends to push up the size of the water heater, but combustion of unused fuel reverses that trend, a not unusual design compromise. Hexis trial units were in domestic operation, sponsored by European utilities. On its listed web site, Sulzer Hexis gave an outline of its initial project, the HXS 1000 PREMIERE, a 1 kWe domestic, grid parallel, fuel cell with supplementary gas firing for larger heat outputs. In Figure 4.9, the fuel cell is in the cylinder. The remainder is the heat exchanger, with control apparatus. The cell details are shown in Figures 4.10 to 4.14.

It should be remembered that the Hexis SOFC was yet another **incomplete** cell, without circulators, and with doubtful published performance assessment in terms of the calorific value of the fuel, in contrast to the chemical exergy.

4.14   ECN/INDEC PETTEN, THE NETHERLANDS

A process is on-going whereby the German firm H C Starck of Goslar, a member of the major Bayer group, is acquiring a large stake in the

Figure 4.9 External view of the Hexis SOFC

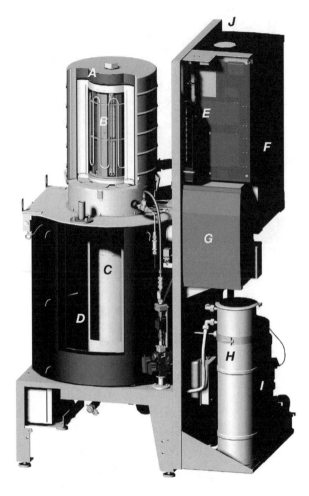

A Thermal insulation
B Stack
C Heat exchanger
D Heat storage tank
E Control
F Auxiliary burner
G DC/AC converter
H Gas desulphurisation unit
I Water treatment
J Exhaust

Figure 4.10   Internals of the Hexis SOFC

Energy Centrum Netherlands (ECN) and in the SOFC/MEA producer
Innovative Dutch Electroceramics (InDEC), jointly sited at Petten. Starck
will acquire SOFC research, operational and manufacturing capability,
and will contribute worldwide marketing strength.

The world's first 100 kW stack was built by Siemens Westinghouse
and operated by Petten utilities. InDec makes MEAs with 5 μm electrolyte, compared with Global Thermoelectric at 6 μm. An anode of
Ni/CGO gave 15% better performance than Ni/YSZ.

Figure 4.11 Flow plate of the Hexis SOFC

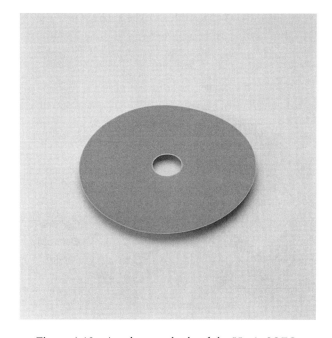

Figure 4.12 Anode or cathode of the Hexis SOFC

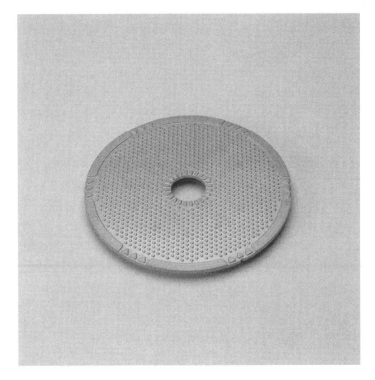

Figure 4.13   Electrolyte of the Hexis SOFC

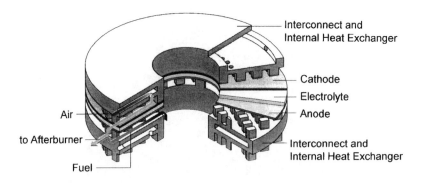

Figure 4.14   Assembled fuel cell of Hexis 1000 Premiere

4.15   REMARKS

The incomplete SOFC is evolving vigorously. There is a diverse stable of high-temperature and intermediate-temperature species. There is
also potential for direct hydrocarbon oxidation, without the need for a
reformer and its irreversibilities. The system, however, may have to face
up to the change of logic described in Figures A.1 and A.2. Isothermal
oxidation may have to be incorporated into the theoretical background
of the industry so that large practical consequences and hence profitabil-
ity can result.

5

Molten Carbonate Fuel Cells (MCFCs)

Control of the passes was, he saw, the key,
To this new district, but who would get it?
W. H. Auden

This quotation gives the flavour of the intense competition among fuel cell types. Survival in the struggle is the main thrust.

5.1 INTRODUCTION TO THE MCFC

The MCFC and, as in Chapter 4, the IT/SOFC are direct competitors for the stationary power role, with convergent operating temperatures. The MCFC is fixed at its electrolyte melting point, but see Ceres Power in Section 4.6. The modus operandi and the general arrangement of an MCFC are shown in Figures 5.1 and 5.2.

The MCFC (Doyon *et al.*, 2003) can use natural gas directly via medium-temperature anode reform, and therefore does not require a commercial and cheap hydrogen source. The MCFC is 'incomplete' in the sense of circulators as covered in the first three chapters and Appendix A.

In Japan, IHI, Mitsubishi and Hitachi are prominent in the MCFC business. Italy (Ansaldo Research) and the Netherlands (ECN-InDEC) are major players in Europe. Korea, via the Korea Institute of Science

Fuel Cells, Engines and Hydrogen – An Exergy Approach   Frederick J. Barclay
© 2006 John Wiley & Sons, Ltd

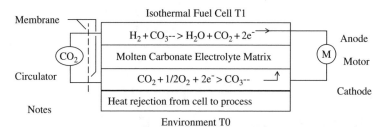

At equilibrium there is zero electric current, and the motor terminal potential difference is the same as that of the cell. The CO_3-- ion circulation is zero. Inside the cell, ion migration in the potential gradient is matched by counter migration in the concentration gradient. (Hydrogen Fuel)

At power CO_2 must be recirculated, anode to cathode, to supply CO_3-- at the cathode to flow through the electolyte, and oxydise the fuel at the anode. The CO_2 is recirculated as anode gas. See Figure 5.4

With methane fuel and a reformer, the port-manteau equation, which embraces reform and shift reactions is:

$$CH_4 + 2H_2O > CO_2 + 4H_2$$

The diagram shows the desired electrochemical process leading to the generation of Vn. The real anode and cathode have additional secondary reactions. Notably at the cathode, there are the peroxide and superoxide paths, Eqns 5.1 and 5.4 in the text.

The anode in the diagram is a porous nickel structure with the pores containing some molten carbonate. The cathode is porous nickel oxide also containing molten carbonate.

Figure 5.1 Modus operandi of the MCFC

and Technology and Korea Electric Power Research Institute, is pursuing a growing systematic programme. The German MAN and US FuelCell Energy, Inc., DirectFuel Cell programmes are collaborative and large (Doyon *et al.*, 2003). FuelCell Energy operated a 1.8 MW prototype DirectFuel Cell in 1996–1997.

The company web sites (see Appendix C) are all useful sources of information, with facilities for e-mail. MAN, due to vigorous growth, has a new web page.

The number of US patents allocated to FuelCell Energy (FCE) is 20 and growing (most recent 2005), an indication that the firm continues to be vigorous. The Price Waterhouse review of the fuel cell industry allocates FCE second place to Ballard in making very large losses of the industry, which raises the question of how long the losses can continue.

5.1.1 MCFCs of FCE and MTU

The 300 kW MCFC of FCE, Danbury, CT, and of the Motoren und Turbinen Union (MTU), Friedrichshafen, has a 600 °C modus operandi as in Figure 5.1. FCE offers larger models in the megawatt range. The

Sketch of MCFC Stack or Battery

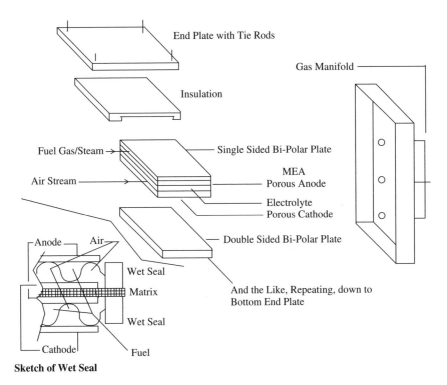

End Plate with Tie Rods

Gas Manifold

Insulation

Fuel Gas/Steam →

Single Sided Bi-Polar Plate

MEA
Porous Anode

Air Stream →

Electrolyte
Porous Cathode

Double Sided Bi-Polar Plate

Anode   Air

Wet Seal

Matrix

And the Like, Repeating, down to
Bottom End Plate

Wet Seal

Cathode   Fuel

Sketch of Wet Seal

Figure 5.2 Arrangement of MCFC

recirculation of CO_2 shown in the figure can lead to easy separation and sequestration of that gas. See the company's web site.

The MTU MCFC provides catalysed 600 °C anode reform capability, with flat anode temperature distribution. In contrast the 1000 °C SOFC encounters difficulties with anode reform, in which excessive reaction rates lead to unacceptable, thermally stressed, local anode cool zones. The title direct fuel cell (DFC) is used, to highlight the absence of a separate combustion-heated 800 °C reformer and its pre-reformer. The balance of plant flow sheet is shown in Figure 5.3.

The resulting hot module is a notable piece of simplification by MTU which leapfrogged the US parent firm. There has now been a double leapfrog via vigorous investment and reorganisation is a new manufacturing site in Germany to meet the needs of a raft of new commercial customers (see web site).

Historic MCFC flow sheets had a pre-reformer for hydrocarbons above CH_4, which the reduced-temperature, MTU 600 °C anode reform

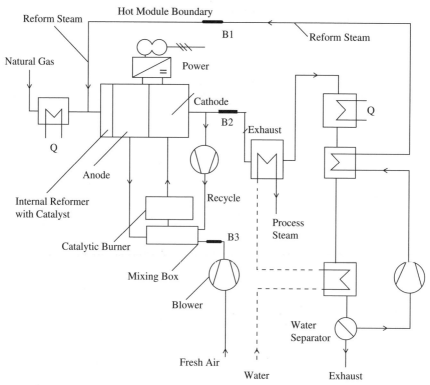

The B2 exhaust fires a steam generator, auxiliary heaters, and water recovery equipment.
The exhaust from methane oxidation is steam laden.

Figure 5.3   Flow sheet of MTU hot module and balance of plant

process made unnecessary. The associated picture of the DFC chemistry
was that pre-reform, reform and shift reactions were locally catalysed in
homogeneous endothermic reactions near the anode, and the resulting
hydrogen consumed in the heterogeneous electrochemical reaction to
yield power, reversible heat, irreversible heat and some unused fuel.
At the anode surface, there is a dynamic equilibrium, in which the
exothermic electrochemical oxidation reaction voraciously promotes the
endothermic reform reaction and swallows the products. The reform
process is invigorated, and its temperature maintained. Reform becomes
very effective at the relatively low 600 °C.

In the MTU flow sheet, preheated natural gas and reformer steam are
supplied to the combined reformer/anode. See Figure 5.4.

After fuel cell isothermal chemical oxidation (Figures 5.3 and 5.4),
the depleted fuel, steam and CO_2 exhaust is partially recirculated to

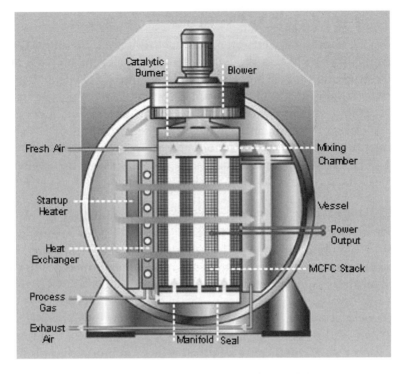

Figure 5.4 MTU illustration of hot module

the mixing box, while the remainder goes to thermal process. Blown cool, fresh air (air heater eliminated) is mixed with the recirculated anode exhaust gas, in the mixing box, the outlet of which supplies the cathode via a catalytic burner. The catalytic burner compensates for the cooling effect of the unheated fresh air. The resulting mixed and reheated flow ensures that the cathode is provided with CO_2, which, with the addition of oxygen, provides the necessary CO_3^{2-} ions. The hot module in Figure 5.4 is a truck-transportable, insulated, fuel-cell-based power plant and (largely combustion-based) heat generator. Initial operating experience is being obtained with this design at 250/400 kWe, with the objective of demonstrating safe continuous stack operation for 40 000 hours at 600 °C, and the achievement of 23 500 h. (Doyon *et al.*, 2003) is unclear on 40 000 h operation on behalf of FCE.

A so far exclusive feature of the competing IT/SOFC is, however, the nascent ability (Chapter 4) to oxidise methane and other hydrocarbon gases electrochemically, without reform. The future MCFC is likely to have to meet this potential leapfrog of its simplicity and fuel versatility.

5.1.2   Detailed Fuel Cell Description

The MCFC membrane electrode assembly (MEA) comprises three layers: a porous lithiated NiO cathode structure and a porous Ni/NiCr alloy anode structure, sandwiching an electrolyte 'matrix' (see detail below). To a first approximation, the porous, p-type semiconductor, nickel oxide cathode structure is compatible with the air oxidant, and a good enough electrical conductor. The nickel anode structure, coated with a granular proprietary reform reaction catalyst, is compatible with natural gas fuel and reforming steam, and is an excellent electrical conductor. As usual, the oxygen is the actual cathode and the fuel the anode. Hence the phrase 'porous electrode structure'.

Conceptually, the wetted electrolyte matrix immobilises the molten carbonate electrolyte, in its pores. (See Section 5.1.8 for unanticipated electrolyte mobility in temperature gradient.) It is manufactured as a sheet of ceramic particles, usually $LiAlO_2$, and typically mixed with powdered alkali metal eutectic carbonates ($62\%Li_2Co_3$, $38\%K_2CO_3$), melting point 488 °C. The difference in thermal expansion (10×10^{-6} °C^{-1}, for particles, twice that for eutectic of the two constituents in the solid state) can lead to cracking in start-up or cyclic operation, and hence to a variety of strengthening additives which minimise that problem, notably CeO. The strengthening additive effect is via the high coefficient of expansion, and melting point low enough to allow sintering with the $LiAlO_2$ ceramic particles. Various matrix recipes are described in Huang *et al.* (1998). In addition to the electrolyte, which fills the matrix to the point of gas impermeability, the pores of the anode and cathode structures are partially filled with electrolyte to create an extended porous three-phase boundary region, leading to increased exchange current

5.1.3   Matrix Initiation

The virgin, or green, electrode structures and matrix have to be matured as part of well-supervised, fuel cell commissioning, procedures. Burnout of organic binders, used in manufacturing, has to be accomplished. Melting of powder electrolyte into matrix and electrode pores, and, in parallel, cathode nickel oxide (NiO) formation, must be achieved. A change of fuel cells after 40 000 h will therefore represent a significant planned outage, for the commissioning of the replacement fuel cell stack.

5.1.4   Matrix and Cathode Deterioration

An aim of the evolving matrix recipes directed at 40 000 h operation is to minimise the slow progressive dissolution of cathode NiO by the matrix, and precipitation of short-circuiting nickel in the matrix, as described in the Japanese paper Morita *et al.* (2001). Korean authors pursue the same objective. They use addition of alkaline earth metal oxide MgO to the NiO cathode to slow down nickel dissolution in less acidic conditions, with demonstrated success.

MTU describes a double-layer cathode: a first layer of lithium-treated NiO and a second layer of cerium-activated lithium cobaltite. The objectives are reduced polarisation resistance and longer life. No performance details are given.

From the foregoing it can be seen that the cathode is once again the more difficult electrode structure. Optimisation is proceeding, in parallel with initial operating experience.

5.1.5 Performance of Complete Cells

The best achievement is that of MAN, namely 23 500 h, as given on its web site (see Appendix C).

5.1.6 Bipolar Plates

The MEAs of the MCFC are sandwiched between stainless steel or nickel bipolar plates, which collect current from, and distribute fuel/steam and oxidant air plus recirculated anode gas via the porous electrode structures to, the chemical reactions at the electrolyte/gas interfaces, Figure 5.1. In the long-anode, narrow-cathode, rectangular MTU fuel cell, fuel utilisation is about 80%, oxygen utilisation about 50%.

5.1.7 Stacks

The clamped stack, Figure 5.2, of MEAs, bipolar plates and end plates which make up the fuel cell battery, has to be sealed at the edges. The technology takes advantage of the molten carbonate in the matrix, and is known as wet sealing. (See sketch in Figure 5.2, bottom left). The bipolar plates have no flow distribution passages at the edges, and the resulting flat surfaces, with anti-corrosion Al_2O_3 coatings, are clamped, both sides, on the MEAs, Figure 5.2. There are, of course, entry and exit passages for fuel/depleted fuel and for air/depleted air (not shown). The thinking which underlies wet sealing is based on isothermal surface tension wetting notions, which break down in the presence of temperature gradients. Surface tension gradients in the presence of temperature gradients are shear stresses, which can move things around. Operating experience being obtained by MAN and described on its web site will establish the practical importance of this notion.

5.1.8   Gas Turbine Integration with an MCFC

Gas turbine/MCFC integration is a complex objective (Morita *et al.*, 2001), which at the small end commences with something like the MTU hot module and one of a current variety of commercial micro gas turbines. Having read about integrated SOFC/gas turbine systems in Chapter 4, the reader will see that the MTU hot module vessel immediately becomes a pressure vessel at the compressor outlet pressure, and enlarges to accommodate the micro turbine. The result is something bigger, and heavier, than the original hot module, and contains very hot combustion chambers. With hindsight, the compact hot module looks attractive!

However, in Williams (2002), FCE authors have rethought the MCFC/gas turbine hybrid problem, and have come up with an alternative system, in which the fuel cell is operated at atmospheric pressure. Pressure occurs only in the gas turbine. Applicability from 250 kW to 40 MW is claimed. The system is still a concept waiting to be realised.

Heated fuel and water are reformed and isothermally oxidised at the MCFC anode. The unused fuel and depleted air from the anode are burnt with added air in a catalytic oxidiser, the output of which heats the cathode and supplies it with oxygen. The cathode exhaust heats the incoming fuel and water in a heat exchanger. The latter exhausts to desired users, for example steam generation or thermal process.

In Morita *et al.* (2001), Japanese researchers from Ishikawajima Harima Heavy Industries (IHI) and the Central Research Institute of the Electric Power Industry (CRIEPI) advocate the need to change the matrix recipe from Li/K to Li/Na, for the increased fuel cell pressure range associated with gas turbines.

An educational way to an improved understanding of the MCFC is to look at the future possibility of putting a large gas turbine in the fuel cell stack exhaust. Power is produced by burning the depleted fuel from the anode, using the hot depleted air from the cathode, in a relatively inefficient gas turbine. Repeating what was said in Chapter 4, the arrangement is emphatically not a bottoming or topping system. The fuel cell is not a heat cycle, and is cooler than the gas turbine, which is a heat cycle. Gas turbine integration is explained in Figure 5.5.

Figure 5.5 is one way of stating the integration problem. In Barclay (1998), the author made an exergy analysis of a combined cycle, combined heat and power installation, based on the real gas turbine of Figure 5.5. It is immediately apparent that the machine needs some alteration to fit a 600 °C MCFC. The operating temperature of the MCFC is related

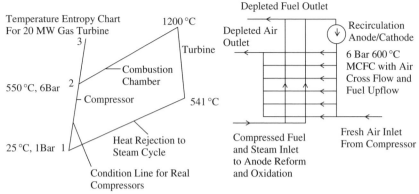

Figure 5.5   Gas turbine integration with an MCFC

to its molten carbonate eutectic, and cannot be adjusted downwards. An upward adjustment would be a major exercise, both for the MCFC, its anode reformer, and for its pressurised containment. On the other hand, the pressure ratio of the gas turbine can be reselected, resulting in a raised compressor outlet pressure and temperature (600 °C). The cell containment can then be designed for the selected pressure. The designer is free to select gas turbine compressors with terminal conditions of pressure and temperature, but only on the Figure 5.5 condition line 123. The slope of this line (a thick line) represents currently achievable compressor efficiency (not to scale). The gas turbine cycle has then to be optimised. The hot pressurised module/fuel cell conditions are determined by the compressor outlet conditions.

For completeness it can be said that integration of MCFCs with Ackeret-Kjeller gas turbine cycles, with twin intercooled compressors, twin turbines and a heat exchanger, is a distant problem, offering new combinations of pressure and temperature for a double fuel cell installation at two pressures. The necessary mass production of fuel cells will wait for research and development, to settle down and the consideration of the design alternatives of the present day.

Also, in the USA and in Williams (2002), pp. 498–501 and 518–521, the Allen Engineering Co., Southbury, CT, is applying its separator, interconnect and flow plate manufacturing technology to the development of a 45 kWe combined heat and power MCFC, another incomplete fuel cell.

5.1.9 Nickel Oxide Deposition at the Cathode at High Pressure

Japanese authors (Morita *et al.*, 2001) found that high pressure up to 45 bar accelerated the dissolution of cathode nickel oxide, and shortened the cell life, via precipitation as nickel in the electrolyte. The problem was ameliorated by using Li/Na rather than Li/K electrolytes. In the Li/Na electrolyte the nickel precipitation was observed to be half that in Li/K, and uniform, rather than in a layer near the cathode. The Li/Na cell, moreover, was capable of $3.3 \, \mathrm{kW \, m^{-2}}$ rather than the $1.2 \, \mathrm{kW \, m^{-2}}$ achieved in the past. That fact will, in the world of patents, cause a stir of appraisal and manoeuvre.

5.1.10 Nickel Behaviour, Short-Circuiting

Morita *et al.* (2001) also examine the mechanics of nickel short-circuiting in Li/Na cells at ambient pressure. There is a process of linking local precipitation sites into chains, which extend into short circuits.

In the light of Morita *et al.* (2001), Figure 5.1 is clearly an oversimplification of the MCFC. At the cathode, a superoxide path (SOP) and a peroxide path (POP) are involved:

$$\text{SOP} \quad 3O_2 + 2CO_3^{2-} = 4O_2^- + 2CO_2 \tag{5.1}$$

$$[O_2^-] = K_1 p_{O_2}^{0.75} p_{CO_2}^{-0.5} \tag{5.2}$$

$$O_2^- + 2CO_2 + 3e^- = 2CO_3^{2-} \quad \text{Superoxide} = O_2^- \tag{5.3}$$

$$\text{POP} \quad O_2 + 2CO_3^{2-} = 2O_2^{2-} + 2CO_2 \tag{5.4}$$

$$[O_2^{2-}] = K_2 \, p_{O_2}^{0.5} \, p_{CO_2}^{-1.0} \tag{5.5}$$

$$O_2^{2-} + 2CO_2 + 2e^- = 2CO_3^{2-} \quad \text{Peroxide} = O_2^- \tag{5.6}$$

5.1.11 MCFC Integration with Coal Gasification

Hotchkiss (2003) presents a timely review of the three main coal gasification processes: fixed bed, fluidised bed and entrained gasifiers. These are aimed at providing gaseous fuel for combined cycle gas/steam turbine installations of large size. The aim of gasifier plants is to heat coal, with restricted air supply, and using a minimum of its own calorific value, so as to produce a maximum supply of gaseous fuel. The analysis of the fuel gas can be adjusted in several ways, according to need: for example, sulphur removal, carbon dioxide removal, hydrogen production by reacting carbon with steam. The British Gas Lurgi fixed bed gasifier employs oxygen separated cryogenically from the air. That would be advantageous if the gaseous fuel were aimed at the future alternative of MCFC or SOFC/combined cycle hybrids. The fuel cells could also use oxygen, and achieve an efficiency gain. Access is given to Texaco information on gasification, 'Introduction to Gasification', by N Richter, at http://www.gasification.org/Docs/O2Richter.pdf/.

At Hatfield in the UK, Jacobsen Consulting is seeking planning permission for a plant which would be based on a Texaco gasifier. Other details are not yet announced. In a 20-year leap forward one could replace the heat engine power plant with a fuel cell/combined cycle hybrid and a membrane-based oxygen enrichment plant (Dyer et al., 2000). The latter reference records the development of a new technology for oxygen enrichment via concentration cells. The principle of operation is as follows. With air at both cell electrodes there is zero potential difference. A potential difference applied between the electrodes will lead to the generation of a concentration difference. The practical case involves flowing air. The system is directly intended for synthesis gas production in gasifiers.

On the one hand, an MCFC with steam and methane input to its reforming anode provides a combustible hydrogen/steam/oxygen/ nitrogen mixture to a gas turbine. Coal gasifiers, on the other hand, one of six competing types, provide a rich hydrogen and carbon monoxide syn-gas as an MCFC fuel. Further integration with a gas turbine, so as to combust the unused fuel, is then inevitable. In larger installations a steam turbine would be added with its heat recovery steam generator. The importance of this technology is a function of the likely future scarcity of natural gas, in contrast to the certain continued availability of coal. The increasing world population is of course very relevant, as is the substantial development potential of the fuel cell, with circulators (see the introduction).

5.2 MCFC STATUS

The vigorous commercial development at MTU in Germany is described on its web site. Initial operating experience with small 250/400 kWe stationary MCFC power units has been gained up to 23 500 h, and as a result there will be evolutionary change. A new manufacturing site has been set up and a raft of commercial customers exists. That makes the MCFC the most prominent existing development of empirical fuel cells. But the basis of customer confidence is a down load of technical data from the company's web site showing efficiency calculations on a calorific value basis. The theory given in this book in the introduction and Appendix A of isothermal oxidation and incomplete fuel cells might affect that confidence by changing the efficiency numbers and bringing about a change of perspective.

The MCFC is now well ahead of the SOFC, but is lower in development potential, in the absence of the capability to oxidise hydrocarbons directly. The SPFC is relatively mature, but for natural gas fuel in the stationary power role it carries the burden of a combustion-fired reformer. The market will answer the question as to whether there can be more than one survivor! If the tripling of MCFC power density, cited in Yoshikawa *et al.* (2001), remains valid, that will be very significant.

5.3   REMARKS

Like the competing IT/SOFC, the MCFC is an incomplete system which has not considered isothermal oxidation with its circulator problem, as defined in this book. The system is also immature as it stands, since the mobility of its electrolyte is a newly recognised fact, for which the author has suggested surface tension gradient as the mechanism. Moreover, a major increase in power density is under consideration, which would interact with electrolyte mobility via higher temperature gradient.

Integration with gas turbines has been thought about but not yet attempted.

The protagonists of the MCFC have a situation well illustrated by the initial Auden quotation, a situation requiring long-term persistent investment in the business of learning from operating experience and reacting to competition and also new theories.

6

Polymer Electrolyte and Direct Methanol Fuel Cells

This, like thy glory, Titan, is to be,
Good, great and joyous, beautiful and free.
                                    Shelley, 1792–1822

6.1 INTRODUCTION

6.1.1 Ballard Power Systems

Ballard Power Systems of Canada (see web site) is the Titan, the co-ordinating major force in the international proton exchange fuel cell or polymer electrolyte fuel cell (PEFC) business. Ballard claims over 500 patents! Also, a worldwide complex of allies serves its international interests.

Although the Ballard PEFC itself is the subject of very thorough materials development on an international basis, and although stack arrangement has been skilfully evolved via operating experience, the system is nevertheless 'doubly incomplete', firstly because a source of cheap hydrogen, the only fuel for the system, is lacking; and secondly, the absence of circulators has not been considered. In Appendix A of this book the author makes a suggestion as to how more efficient and cheaper hydrogen manufacture might be achieved. That is a future, time-consuming, development problem. As a result the industry may move slowly. Commercialisation is far from imminent in the author's opinion.

Fuel Cells, Engines and Hydrogen – An Exergy Approach   Frederick J. Barclay
© 2006 John Wiley & Sons, Ltd

Moreover, hydrogen storage onboard vehicles and ships remains in pressurised tanks. An alternative solution seems to be needed, which may be the liquid hydrogen tank developed by Air Liquide for the Renault fuel cell. See the listed Renault web site for the Renault/du Vera vehicle fuel cell project.

A convenient tactic for this book is to say what Ballard is doing to design, build and sell its stacks, and to compare and contrast the tactics of the growing number of latecomer competitors and their technologies. Such competitors do not have access to restricted Ballard patents such as flexible graphite. Information on competing technologies is not plentiful! Patents are invaluable. In the references many patents are included which are not dealt with in the text. That is because they represent further reading for a very wide spread of readers, who are saved the labour of patent searching for their topics.

The Ballard web site has a road map pointing to the achievement of US$30 per kW by 2010, via improved power density, stack cost, stack durability, freeze start capability, better water management, and the like. All of that is underlain by interactions with Johnson Matthey UK on MEA design, by co-operation with Victrex UK on electrolyte development (alternatives to Nafion), and by co-operation with Graftech on flexible graphite and its potential for very cheap flow plates and sophisticated cell design. The latter two activities, exclusive to Ballard, greatly strengthen the Ballard competitive position. A notable omission is that there is no mention on the Ballard web site of neighbouring methanol focused Methanex and the need for the development of a hydrogen mine.

6.1.2   Ballard History

PEMFC (Proton Exchange Membrane Fuel Cell) and SPFC (Solid Polymer Fuel Cell) are the two competing mnemonics of a low-temperature fuel cell type originated for use in space by General Electric, USA. To reflect present practice, the author will use PEFC (Proton Exchange Fuel Cell). The DMFC (Direct Methanol Fuel Cell) also uses proton exchange membranes, but is referred to by its own mnemonic. Proton exchange between polar water molecules is discussed by Koryta (1991; 1993) and in the introduction to this book.

Ballard is now big enough to acquire competitors sponsoring alternative types of fuel cell, in the possible event that Ballard's early bet on the PEFC proves not to be 'on the nose'. Its fortunate PEFC bet (Prater, 1990) was made when platinum utilisation was very poor, and present successful developments, notably flexible graphite flow plates, were not foreseeable.

In the same way, the result of competition amongst the three surviving PEFC, SOFC and MCFC fuel cell types is not predictable. For example, the SOFC has the nascent ability to oxidise natural gas directly, and the MCFC is fuel omnivorous as a result of its mature 600 °C isothermal anode reform capability. Those latter attributes are in contrast to the confinement of the PEFC to hydrogen of minimal CO content, from hydrocarbons processed in an inefficient combustion-driven reformer (inefficient relative to anode reform). The Ballard PEFC has, however, achieved high power density with good, but not unlimited, manoeuvrability.

6.1.3   Ballard Status

Ballard is a worldwide organisation with large shareholders, namely Ford, General Motors, Daimler-Benz (Daimler Chrysler) and, on the power side, GPU International and EBARA in Japan. In addition to stationary power stacks of 250 kW, EBARA is pushing hard on a 1 kW domestic PEFC (Ballard Press Release BWO117, 22 February 2001, via web site businesswire.com).

Honda uses Ballard Mark 902 cells in it FCX demonstration 78 kW fuel cell vehicle, which has power storage in an ultra-capacitor of Honda manufacture (see the 2002 press release 31247/24072 as listed on the Honda web site).

Details of all Ballard experimental vehicles are to be found on its listed web site, in the Media Resources section, Image Gallery.

6.1.4   Ballard Stacks

A Ballard stack is shown in the expanded stack of Figure 6.1.

6.1.5   Flexible Graphite and Ballard

The 30-year technological story of flexible graphite, and example modern flow plate/fuel cell descriptions, are given in four Graftech patents (Mercuri, 2001a, b; 2002a, b).

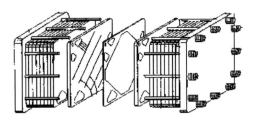

Figure 6.1   Exploded Ballard stack with flow plate and membrances

Flexible graphite is a form of graphite in which the distance c between the basal planes $(a \times b)$ of carbon atom hexagons has been expanded, 80 times or more, by strong acid chemical treatment (intercalation). The grossly expanded graphite is at first in the form of worm-like or vermiform particles, which can be partially re-compressed into web, paper, sheet and board materials for a great range of applications, benefiting from flexibility and from anisotropic thermal, electrical and diffusion properties. For Grafcell, the grade of flexible graphite for fuel cells, Advanced Energy Systems Inc. has developed perforation and groove creation techniques appropriate for fuel cell electrodes and flow plates. Such 'machining' or 'forming' modifies the local anisotropic properties, and advantage can sometimes be taken of that fact. The listed AES/Graftech web page asserts that flexible graphite flow plates are **ten times cheaper** to manufacture than previous natural graphite articles, a big economic advantage for Ballard with its exclusive access to Grafcell. Grafcell flow plates are used in the Ballard Mark 900/902 stack, as used in the Daimler Chrysler Necar 5 and Jeep Commander projects. (See the Ballard web site.)

In Mercuri *et al.* (2001a, b) Graftech proposed an evolved notional fuel cell design, which tackles the topic of even less platinum utilisation, via a printing pattern for platinum, as opposed to an even spread. See Figure 6.2.

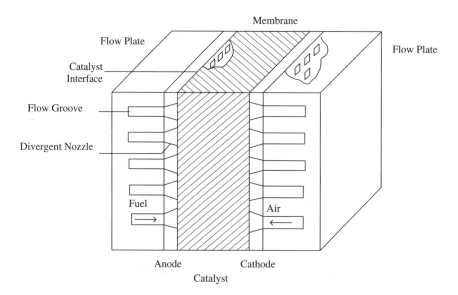

Figure 6.2 Notional Graftech fuel cell arrangement

The cell current prefers to flow at the periphery of the perforations, where the punching process has increased the local electrical conductivity. The platinum is located close to the current peaks.

The early evidence on Ballard thinking about flexible graphite is presented by Wilkinson *et al.* (1996a, b). The latter patents are offspring of the 1994 '370 Patent', well known in the PEFC business (Washington *et al.*, 1994), and which sets out alternatives in flow plate design. See Figure 6.3.

For the Ballard Mark 900/902 stacks (Gibb, 2001) using Grafcell flow plates, a meeting of minds must have occurred between theAES/Graftech

Figure 6.3 Flow plate design alternatives from US Patent 5,527,363

thinking underlying Figure 6.1 and that in Mercuri *et al.* (2001b), Ballard experience behind Figure 6.2 with Wilkinson *et al.* (2000a, b, c) and also with Johnson Matthey thinking on platinum deployment. The details are not known. However, Figure 6.2 demonstrates that Johnson Matthey/Ballard, these days, can make up and try out experimental fuel cells in the laboratory with relative ease, and check voltage–current characteristics.

Notional Grafcell flow plates are shown in Figure 6.4 by courtesy of Advanced Energy Products Inc. The actual design of the Ballard flow plate is not given. The arrangement is shown of a Ballard PEM fuel cell using flexible graphite plates.

6.1.6 Ballard MEAs

Ballard MEAs are supplied by Johnson Matthey UK from a single-purpose new plant in the UK, backed by research laboratories. The Ballard MEA technological story is given by Ralph *et al.*, 1997, followed by Ralph and Hogarth (2002) and Ralph *et al.* (2002; 2003), and which are paralleled in India by Kumar *et al.* (1995). The battle has been to get

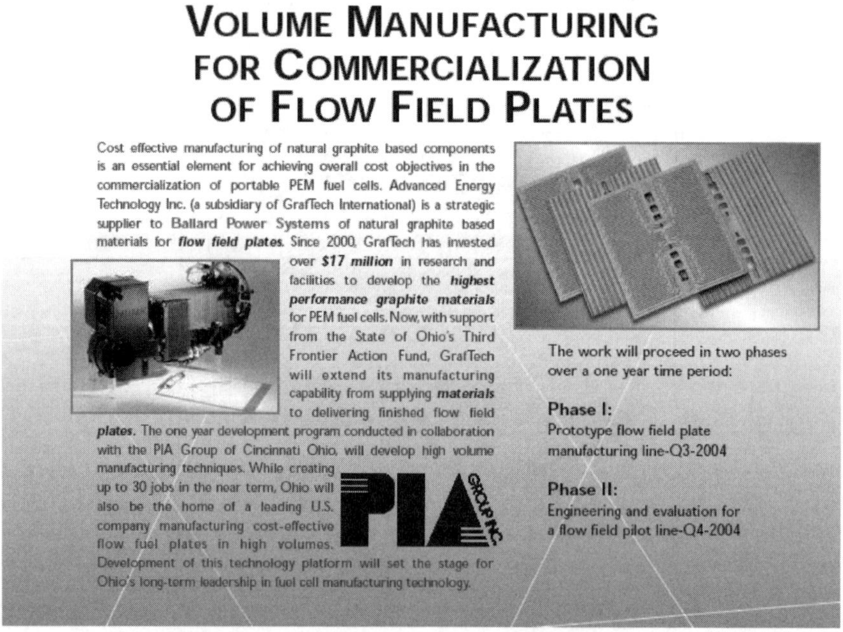

VOLUME MANUFACTURING FOR COMMERCIALIZATION OF FLOW FIELD PLATES

Cost effective manufacturing of natural graphite based components is an essential element for achieving overall cost objectives in the commercialization of portable PEM fuel cells. Advanced Energy Technology Inc. (a subsidiary of GrafTech International) is a strategic supplier to Ballard Power Systems of natural graphite based materials for **flow field plates**. Since 2000, GrafTech has invested over **$17 million** in research and facilities to develop the **highest performance graphite materials** for PEM fuel cells. Now, with support from the State of Ohio's Third Frontier Action Fund, GrafTech will extend its manufacturing capability from supplying **materials** to delivering finished flow field **plates**. The one year development program conducted in collaboration with the PIA Group of Cincinnati Ohio, will develop high volume manufacturing techniques. While creating up to 30 jobs in the near term, Ohio will also be the home of a leading U.S. company manufacturing cost-effective flow fuel plates in high volumes. Development of this technology platform will set the stage for Ohio's long-term leadership in fuel cell manufacturing technology.

The work will proceed in two phases over a one year time period:

Phase I:
Prototype flow field plate manufacturing line-Q3-2004

Phase II:
Engineering and evaluation for a flow field pilot line-Q4-2004

Figure 6.4   Flexible graphite flow plates

down to frugal utilisation of platinum catalyst, based on initial progress at Los Alamos and at Texas A & M University. Latterly, flexible graphite technology has appeared, which is of low cost and has the potential to yield further reduction of platinum loading. It is not clear whether this possibility has been used by Ballard.

Graftech summarises the properties of Grafcell as corrosion free, low specific density, low contact resistance, high electrical conductivity, impact resistant, high thermal conductivity, recyclable, cost-effective manufacturing and thin.

6.1.7   Nafion and Alternatives

The classical DuPont proton exchange electrolyte is Nafion. The formulation, given in Lovell and Page (1997) is

$$[-(CF_2)_m-CF-CF_2-]_n$$
$$|$$
$$O$$
$$|$$
$$CF_2$$
$$|$$
$$CF_3-CF-O-(CF_2)_2-SO_3^-H^+$$

A sulphonic acid radical is at lower right. Its H^+ proton can gather a hydration shield from water in the membrane, the simplest version being H_3O^+, the hydrated proton. In Lovell and Page (1997) the figure of merit for a solid polymer electrolyte is defined as the equivalent weight (EW), the polymer mass in grams containing 1 mole of sulphonic acid, the latter being the ionic agent of proton conductivity (Koryta, 1993). Nafion 112 is given the value $EW = 1100$, at the top of the range. A convenient mental picture of an ionomer comes from considering it as a pebble bed. The pebbles are stuck to each other where they touch: a coherent open-pored solid, which must have water in its pores to facilitate proton exchange, involving the fixed position, sulphonic acid radicals on the surface of the pebbles.

Williams (2002), Chairman of E I Du Pont de Nemours, describes its new process (solution cast membrane technology) for manufacturing Nafion membranes aimed at PEFC/MEAs. With large production the membrane NR112 can fall to about $80 per m^2, and NR111 to about $45 per m^2. Details of the two types are given.

Ballard itself, via Ballard Advanced Materials (Lovell and Page, 1997), has been generating a possible Nafion alternative, namely perfluoro-polyvinyl ether sulphonate (BAM3G), with 3G denoting third generation. There is strong Japanese competition, namely from Asahi Chemical Industry, Aciplex (Williams, 2002, pp 62–65). Other competitors named in the latter reference are Dow Chemical, Asahi Glass (Flemion), ICI plc, W L Gore and Associates (Gore Select), Hoechst AG, Hoechst Celanese, Allied Signal, DAIS Corp, etc. Lovell and Page (1997) were prescient in identifying the benefit to be accrued from examples like the Victrex–Ballard alliance, announced in a Ballard News Release dated 22 January 2001. Ballard has established a partnership with the UK company Victrex plc (see web site) for a four-year parallel development programme, involving both BAM3G ionomer, and a well-established Victrex ionomer (PEEK). The mnemonic expands to PolyEther-EtherKetone. The repeat unit is

[-oxy-1.4-phenylene,-oxy-1.4-phenylene-carbonyl-1.4-phenylene]$_n$

The chemical properties, given in Lovell and Page (1997) include solubility in sulphuric acid, and corrosion by benzene–sulphonic acid. Those latter are prima facie evidence that PEEK can be developed into a solid polymer electrolyte equivalent to Nafion 112 for the PEMFC, and Nafion 117 for the DMFC.

The Far East alternatives to Nafion, namely the Aciplex (Asahi Chemical Industry), Flemion (Asahi Glass Company) and Shanghai (Shanghai Institute of Organic Chemistry) membranes, are compared in an experimental cell with varieties of Nafion in Xuezhong Du *et al.*, 2001). These alternatives have not caught on, except with their originators, who are nevertheless powerful competitive organisations. The Chinese entry into advanced technology is a welcome surprise, and their market is obviously very important. The author has presented in the introduction an explanation (readjustment from equilibrium) of the initial dip of the voltage–current curve, which differs from the historic activation complex notion used in Xuezhong Du *et al.* (2001). That notion, however, does not lead to the reversible equilibrium achieved by Bacon's cell.

6.1.8   Alternative Flow Plate Materials Used by Competitors

Alternative materials for bipolar plates include graphite, stainless steels, titanium and aluminium, all with a developed fabrication technique, and coating technique if needed. Major competitors UTC Fuel Cells has an active fuel cell bus programme, but give sparse details of its flow plate and other technology. (See UTC web site.)

Another competitor is Nuvera, which has a PEMFC fuel cell design, 'Andromeda', featuring non-coated metal flow plates (metal unstated). The cell is available to customers who can integrate it into their projects. A major example of the latter is Renault which, with Du Vera, has a large project 'Scenic', aimed at having vehicles on the road in 2007. A range of 500 km is based on a liquid hydrogen fuel, the tank being due to Air Liquide. The Andromeda fuel cell generates 42 V DC, converted to 250 V AC and supplied to a specially developed, highly efficient squirrel cage motor.

ECN Holland works on both the SOFC and PEMFC. For the PEMFC its web site asserts the imminence of breakthroughs in materials development. ECN is active in electrolyte development, electrode and catalyst development, and stack development. No description is given of the detail design features.

When Ballard chose Grafcell flow plates, its agreement with Graftech deferred any opportunity for competitors to use the new material. Some competitors using titanium with proprietary coatings are the Renault/Nuvera/3M's automobile application, and Intelligent Energy, power for the British foot-soldier. Intelligent Energy has provided the design basis for 10 kWe stacks by Innogy Ltd, UK, with photochemically machined titanium flow plates having a proprietary coating by ICI Ltd who assert an improvement (now history) in voltage–current characteristic due to the change from stainless steel to coated titanium ICI, surprisingly, does not mention its flexible graphite competitor, which offers ×10 cheapness (Wilkinson *et al.*, 1996a) and platinum economy.

6.1.9 Ballard Operating Experience

Ballard assembles and tests stacks, to confirm their characteristics and optimise such water management problems as electrolyte humidity control and cathode water removal, via wettable materials in the porous electrode, to the external air flow. Potential difficulties, exclusive to the water-producing PEFC, can be seen with 100% humidity in the tropics, and with freezing conditions in cool climates. High altitudes can be difficult for all fuel cell types, via low oxygen density.

Water management is dealt with by Voss *et al.* (1993; 1995). It is shown that the cell voltage–current characteristic can be greatly improved, and the achievable power density increased, with improved water management. The technique is to create a steeper cathode to anode water concentration profile through the cell. The water at the anode is removed by the fuel gas stream, as a result of inducing a pressure drop in the flow plate channel, which can be optimised for best performance.

Water removal is obviously difficult if the cell needs to operate at zero relative humidity (saturation conditions). The difficulty is eased by the modern trend to increased pressure and 150 °C temperature. See Section 1.9. The result is a dis-equilibrium at the hot exhaust, which facilitates water dispersal.

Ballard supplies stacks to its end-use partners to gain experience with vehicle and stationary power systems. The stacks, and their associated systems, are, and must be, tested in a breadth of operational exigencies.

A current operational problem is damaging anode voltage reversal following fuel starvation. A flow of patents covers the topic (Knights *et al.*, 2000; 2001a; Colbow *et al.*, 2001; Taylor *et al.*, 2001). The problem, relevant to both the PEFC and the DMFC, and to all manufacturers, is discussed by Johnson Matthey in Ralph *et al.* (2003).

Stacks which have seen vehicle service may have been taken beyond their limits of manoeuvrability. As a result, damaging local fuel or air starvation has occurred at the modified cell potential differences, and put some cells in difficulties. Current through the stack continues, and at the air-starved cathodes, hydrogen and heat production ensue. At the fuel-starved anodes, the damage is more severe. Water electrolysis, leading to carbon corrosion, occurs. Gascoyne *et al.* (2001) proposes incorporating in the MEA a sacrificial layer of activated carbon, which would corrode in preference to graphite components (hopefully including flexible graphite). The story is, perhaps, not yet at an end. For example, the ultra-capacitor deployed by Honda (see Section 6.1.3 above) could drastically limit the demands made on the manoeuvrability of the fuel cell stack, and so cut the probability of voltage reversal..

No Ballard competitor mentions voltage reversal, presumably due to inadequate test experience.

There is a push upwards to PEFC operation at around 150 °C to minimise catalyst sensitivity to CO and CO_2 in the reformer-derived hydrogen fuel.

6.2   ELECTROCATALYSIS IN THE SPFC

There is reason to pause here, and refer back to the electrocatalysis discussion in the introduction. Best electrocatalysis yields maximum power density from minimum platinum loading. For a cathode platinum particle to be electrocatalytically effective, it must be in electrical contact with the conducting porous electrode structure. It must also be in contact with the membrane, or a biconductor layer on the membrane surface,

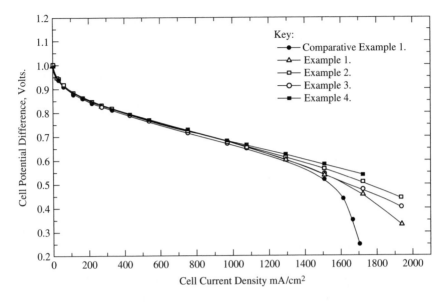

Figure 6.5 SPFC voltage–current characteristic

and via a thin meniscus to which oxygen must have access, to find protons crossing from the anode. In Masahiro *et al.* (1998), it is made clear that a platinum particle in the membrane, but electrically isolated, catalyses the non-electrochemical oxidation of available hydrogen to generate water and heat, but not electrical power. Such water can, however, be used, via titanium oxide particles, for ionomer humidity control (Masahiro *et al.*, 1998). Ballard does not seem to use the latter technology, which is patented.

In Brown *et al.* (2000) Johnson Matthey pursues via five experimental fuel cells, the topic of optimising platinum particle coupling to cathode electrode structure and membrane, while simultaneously improving water removal capability. The resulting cell performances are illustrated in Figure 6.5, reproduced from Brown *et al.* (2000).

Anode catalysts deal with a lesser overpotential problem, but need to be relatively insensitive to fuel-borne carbon monoxide.

6.3 CATHODE VOLTAGE LOSSES IN THE PEFC

Figure 6.5 illustrates well the known shortfall of V_n in the PEFC from 1.2 V to 1 V. Johnson Matthey is beginning to say (Ralph *et al.*, 2002) that there may be an approach to this problem via alternative, or modified, catalyst technology.

6.4   THE PEFC HYDROGEN ECONOMY IN ICELAND

The PEFC fuel supply must be hydrogen from natural gas via a reformer on terra firma or on the vehicle. An obtrusive example is the much discussed proposal for Iceland to have a hydrogen economy, based on cheap geothermal power, and resulting low-cost, CO-free, electrolytic hydrogen for vehicle and ship propulsion. Cheap hydrogen is the *sine qua non* of the hydrogen economy. So far, based on very cheap geothermal and hydroelectric power, only Iceland can fulfil the need, and invigorate PEFC utilisation. With dearer hydrogen, in the international context, pressure is generated on a difficult problem, namely the PEFC electrolyser and fuel cell combined efficiency (Chapter 3). For electrolytic hydrogen from dear power, good electrolyser efficiency is an economic necessity. For hydrogen made by reform from natural gas, with methanol as a parallel product, cheapness via efficiency is desirable, but not easy to achieve. The Regenesys power storage system of Chapter 2 would not have been required if an efficient fuel cell/electrolyser pair were available.

6.5   FUEL SUPPLY

Natural gas is the main source of hydrogen. See Barclay (1998) for details of an early ICI large natural gas reform plant, outputting hydrogen and methanol. Balanced marketing of both hydrogen and methanol from such a plant or a modern Canadian equivalent (Methanex Inc.; see web site) would be needed. The DMFC is therefore a potentially important, but relatively immature, type of PEFC, and, logically, Ballard and Johnson Matthey are active in that area.

Pressurised cylinders are the usual mobile hydrogen store, albeit much effort is being expended on alternatives: see the Renault/du Vera/Scenic web site, and the Air Liquide development of a liquid hydrogen tank.

Whereas the PEFC is a mature device with a doubtful fuel infrastructure, the DMFC is an immature cell with a near mature fuel infrastructure, ready and waiting.

6.6   DMFCs

Methanol is a liquid which could have an infrastructure parallel to that of gasoline. There could be a methanol pump at the local petrol station.

For that reason, and because the system does not need a reformer, the DMFC, currently a system of relatively low power density, has its vigorous protagonists, notably Methanex which would supply the methanol fuel.

In Appendix A, the author did not calculate, using a modified equilibrium diagram, the fuel chemical exergy of methanol. An assault on the calculation would involve the JANAF thermochemical tables (Chase *et al.*, 1998). In view of the immature status of the DMFC, the author deferred the latter significant task. Note, however, that the net enthalpy of combustion is given in Kotas (1995), Table B1, as $726.6\,kJ\,mol^{-1}$, which is of course substantially smaller than the fuel chemical exergy in $kW\,s\,mol^{-1}$.

The work reported by Ralph *et al.* (2003) is a well-rounded, self-contained essay on the DMFC. (See DMFC flow sheet in Figure 6.6.) Moreover, because Ballard/Johnson Matthey did not contribute on fuel cells at the Palm Springs Fuel Cell Seminar in 2002 (see below), Ralph *et al.* (2003) is the current information source, additional to the patents in the list of references. Note that the methanol–water mixture presents to the fuel electrode its associated methanol vapour pressure. The DMFC does not have an incompressible fuel. The cell needs circulators. It is incomplete.

A collection of historic DMFC references, dating back to 1965 (Williams, 2002), added little to Ralph *et al.* (2003), although it did highlight some seminal sources.

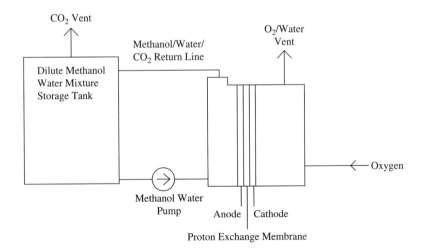

Figure 6.6   DMFC flow sheet

The present-day DMFC is based on MEAs resembling those of the PEFC, but with relatively high catalyst loading and thick Nafion 117 instead of thin Nafion 112. A unique feature is that the liquid fuel is mixed with water, the mixture being recirculated for transport purposes. The fuel/water store is bulky. The methanol is then separated from the stream passing the cell anode, where oxidation occurs. The cell begins therefore by incurring an immediate irreversibility, a mixing loss followed by a separation loss. There has already been irreversibility in fuel manufacture, say at the Methanex plant.

The water ensures that membrane humidity is ample, but also ensures that some miscible methanol crosses to the cathode via the membrane. The MEA is optimised around minimum methanol crossover (Ren X *et al.* 2000):

At the fuel anode $CH_3OH + H_2O = CO_2 + 6H^+ + 6e^-$ $\quad V_n = 0.046\,V$

At the air cathode $\frac{3}{2}O_2 + 6H^+ + 6e^- = 3H_2O$ $\quad\quad\quad\quad V_n = 1.23\,V$

Overall reaction $CH_3OH + H_2O + \frac{3}{2}O_2 = CO_2 + 3H_2O$ $\quad V_n = 1.18\,V$

At the Palm Springs Fuel Cell Seminar (Williams, 2002), the Jet Propulsion Laboratory abstract records the laboratory use in a four-cell stack of a new membrane which cuts the methanol crossover by a factor of 4. The same paper records the demonstration by the Jet Propulsion Laboratory at Palm Springs of a 1 kW DMFC.

Also at Palm Springs, two papers by the Methanol Fuel Cell Alliance (Ballard/BASF/BPAmoco/Daimler Chrysler/Methanex/Statoil) and the Methanol Institute, respectively, portray the existing substantial methanol production, distribution and trading based on natural gas reform/synthesis gas/methanol, as in the Methanex Canada plant. Methanol from biomass is a future possibility. A methanol pump can be fitted within the footprint of many existing diesel/gasoline filling stations, and an 'Identic' refuelling nozzle has been developed in Sweden, to avoid confusion between methanol and alternative fuels.

United Technologies Fuel Cells is engaged in DMFC development, in competition with Ballard/Johnson Matthey. It is a part in the project by Renault to develop the 'Scenic' vehicle fuel cell. Neither for its PEFC, nor for its DMFC (and MCFC), does UTC Fuel Cells offer product-coloured illustrations. Moreover, its literature or listed web site does not deal with the cell voltage reversal problem, mentioned in Ballard patents above in connection with fuel cell bus operation. Accordingly it is not possible for the author to portray the UTC Fuel Cells scheme of things.

The German Julich Laboratories discussed DMFCs of up to 2.5 kW. The water of permeation and, relatively small, the water from fuel oxidation are both collected and recirculated: the slogan is water autonomy. The same objective is pursued in the 10 kW cell of the University of California (Davis).

In the Juelich paper, the use of a semi-permeable membrane for water is mentioned, one of the components needed for the equilibrium diagram of Appendix A. See also Dyer *et al.*, (2000) for an oxygen semi-permeable membrane!

Other papers at the Palm Springs Seminar are concerned with underlying problems, such as better catalysts, improved water management, pressurised operation at 150 °C, and the like, all confirming the immaturity of the DMFC and the big effort being made to mature the system.

Ralph *et al.* (2003) give, at Figure 2, a complex methanol electro-oxidation process at a platinum particle, which highlights the extent of development problems. The same topic was discussed at Palm Springs by University of Washington authors.

DuPont has entered the market for DMFC MEAs, as announced in the final abstract of the Palm Springs Seminar.

6.7 TOKYO GAS COMPANY, DESULPHURISER

The company is partnering Ebara-Ballard with the commercialisation of its 1 kW domestic PEFC, and has developed a desulphuriser in a transparent container. The reagent changes colour as it is used, so that the user can install fresh reagent in due time (see the web site).

6.8 REMARKS

The worldwide fuel cell bus demonstrations based on electrolyser-generated hydrogen highlight the prowess of the Ballard PEFC, and simultaneously demonstrate its untenable long-term position. The pollution at the power plant stack supplying the electrolyser exceeds that removed by the pollution-free exhaust of the bus.

Efficiently produced and cheap hydrogen from a hydrogen mine is a prerequisite to long-term success, in the auther's view. The hydrogen mine project of Figure A.4 involves complete fuel cells based on isothermal oxidation using circulators as in Figure A.1. The Ballard PEFC may have to examine these large issues with its hydrogen manufacturers.

7

Fuel Cell Economics
and Prognosis

Up and down the City Road,
In and out the Eagle,
That's the way the money goes -
Pop goes the weasel!
    W. R. Mandale, 19th century

7.1 OPENING REMARKS

This book set out to define the concept of isothermal oxidation as it
occurs in fuel cells, and to differentiate that form of oxidation from
combustion which does not occur in fuel cells. Whereas combustion and
heat engines are Carnot limited, isothermal oxidation has its own limitation, namely the fuel chemical exergy. The latter concept is based on
the equilibrium diagram, Figure A.1 of Appendix A. The performance,
and therefore the economics of fuel cells, will never be correctly defined
if the concept of fuel chemical exergy is not widely understood. In a
very recent review of the fuel cell industry (Price Waterhouse Coopers,
2005), the industry is allocated the status of being unprofitable, and of
making heavy losses. In such circumstances it does not yet even aspire
to having detailed economics, as does a mature, competitive and profitable industry. One can, however, contemplate the technical status of
existing stationary and vehicular power sources (a moving target), and

assess the chances of drawing level or surpassing them using complete fuel cells, when isothermal oxidation is realised, and made practical. The book makes it clear that drawing level and surpassing is a distant prospect. That point is emphasised by Eaves *et al.* (2004) who compare battery electric and fuel cell vehicles in terms of cost, energy efficiency, weight and volume. The comparison heavily favours the battery electric vehicle.

Complete fuel cells are engineered for isothermal oxidation by the addition of perm-selective membranes and isothermal concentration cells. They would, if developed, generate much higher Nernst potential difference than an existing incomplete fuel cell. That would give a chance for fuel cells to draw level, a chance the industry does not seem to understand.

7.2 FUEL CELL ECONOMICS – SELECTED SUMMARIES

The only available complete fuel cells with promising economics are the power storage devices of Regenesys, described in Chapter 2. The prices, capital cost and operating cost are not available from the new owners, VRB Power Systems of Vancouver. The large pilot installations which were built in the UK and USA were thermodynamically superior, arising from their use of incompressible liquid products and reactants, and visibly modest in materials of construction. There is no competition. The remaining systems such as the hydrogen economy, which are incomplete, can have the same destructive adjective applied to their economics! When systems are incomplete they are also illogical, and without a reasoned thermodynamic modus operandi with which to underpin their economics. With the achievement of completeness capital costs per cell are bound to rise, but overall economics, cost per kilowatt, it is hoped by the author, will improve with improved performance.

In Donelson *et al.* (1998) Ceramic Fuel Cells Ltd, of Australia, gave itself a target of A$1500/kW for SOFC stationary power installations of 200 kW. A rough translation is £600/kW, or US$ 960/kW of incomplete cell.

In Williams, 2002, the Edison Power Research Institute surveys the US market for all cell types, except Regenesys. High capital costs snuff out the phosphoric acid cell, reinforcing the author's decision to cut a chapter from this book on that fuel cell type. For the SOFC the EPRI neatly brackets the Australian estimate. Because the technical underlay to these figures is volatile, their survival capability is doubtful. Relative efficiencies are based on the fuel calorific value, which grossly underestimates the long-term development potential of the fuel cell industry (Chapter 1), and creates huge doubt about the predictions.

In a much more optimistic article in (Williams, 2002), the Solid State Energy Conversion Alliance (SECA) predicts $400/kW from versatile 3–10 kW SOFCs from Siemens Westinghouse or Cummins Power/McDermott Technology. Vehicle applications of the SOFC are included in a wide range of market values.

Further, General Electric gives $388/kW as its SOFC projected cost. Similarly, the PEFC is at $1000–1500/kW, and the MCFC is at $1250–1715/kW. The MCFC still must demonstrate 40 000 h endurance, possibly at high power density.

For their vehicle PEFC named 'Scenic', Renault–Nuvera, commencing with their 'Andromeda PEFC', aim at $30/kW by 2010. The Scenic fuel cell is not likely to compete with more advanced Ballard cells based on cheap flexible graphite rather than on costly titanium flow plates. Whatever technical performance Renault achieves, Ballard can beat it. The lather's patent position is more or less impregnable.

The SOFC may be drawing ahead, but there is plenty of time for system setbacks, and alternatively for leapfrogs' to occur.

The author's view is that new candidates for the list of cells to be reviewed cannot be admitted until the first one shows up with circulators (concentration cells). That, according to the foreword, will be some years after the publication of this book.

7.3   NON-FUEL-CELL MOTOR VEHICLE ECONOMICS

In the *IEE Journal*, January 2003, UK consulting engineers Ricardo Ltd, via Richard Gordon, describe their i-Mo-Gen prototype, 'shallow hybrid' small automobile. The vehicle is driven by a newly developed 1.2 litre diesel engine, allied to a 9.6 kW NiMH battery of 620 W h, at 42 V. A single electrical machine functions as starter motor/charger/drive motor. The diesel engine is reinforced by the electrical machine at the low-power end of its characteristic where torque is low. The shallow hybrid replaces a normal 2 litre engine. The result is a 28% improvement in overall fuel consumption. A large market is foreseen for the next 20 years, without fear of fuel cell competition from the existing incomplete systems. The emergence of complete fuel cells could greatly alter that competitive market.

The Ricardo conclusion is not out of line with a US paper (Williams, 2002) by TIAX LLC of Cupertino, CA, which omits to give its full name, but is a successor to the Technology and Innovation Business of A D Little. The paper concludes that (incomplete) hydrogen fuel cell vehicles will have a $1–2000 per annum operating cost penalty (from high capital cost), relative to engine-driven vehicles.

In the economics-insensitive niche market of the luxury automobile, the development of a high-output SOFC alternative to the lead–acid battery and engine-driven alternator is well advanced at BMI/Delphi Corp (Williams, 2002). BMW is a partner in the project, which uses a Global Thermoelectric planar SOFC stack. The gain would be air-conditioner operation independent of the vehicle engine.

The DMFC is showing promise as a portable power source, but the bulky fuel is a problem (see several papers in Williams, 2002). Vehicle application of the DMFC is some way away. The advent of better ways of processing natural gas, for example the hydrogen mine, will cheapen methanol substantially, and move the DMFC towards a competitive position with the PEFC, which it does not yet occupy. There are no DMFC buses being demonstrated, and no DMFC stationary power plants for sale.

7.4   PRICE WATERHOUSE FUEL CELL INDUSTRY SURVEY

In the IMechE magazine *Professional Engineering* (Vol. 18, No. 19, p. 8) is the headline 'Fuel-cell Developers see Their Losses Mount'. The background is a survey of the industry by Price Waterhouse accountants, which contains the following significant statements:

1. All of the companies surveyed continued to operate at loss as they sought to commercialise.

2. Of the companies surveyed 17 out of 20 were from North America and Canada, split almost evenly between Canada and the USA.

3. Of the four new companies included in the survey this year, two (Ceramic Fuel Cells and Ceres Power Holdings) are focused on developing SOFCs for stationary market applications, and two (ITM Power and QuestAir Technology) are concerned with

hydrogen infrastructure. (Note that ITM Power and QuestAir Technology are in pursuit of economies of fuel cell/electrolyser manufacture economies (ITM) and pure hydrogen manufacture (QuestAir)).

4. The residential cogeneration market in Japan is recognised as one of the most promising near term commercialisation opportunities in stationary power.

5. In 2004 none of the companies surveyed were profitable, a trend consistent since this survey began three years ago. Moreover, the total loss increased from $387 million in 2003 to $465 million in 2004, a movement of 20% year on year.

Ballard Power Systems and FuelCell Energy (Doyon *et al.*, 2003) were the top two revenue (not profit) earners of 2004. (Note, incidentally, that the spellings FuelCell Energy and QuestAir are essential to get information online from the likes of the US Patent Office.)

The big investment is in the PEMFC, but Ceres Power in the UK is using its 500 °C solid oxide cell to enter many applications (Bance *et al.*, 2004).

The industry is developing strategic alliances to help bear the burden of the losses.

7.5 REMARKS

The above discussion covers a situation in which the underlying technology of the fuel cell industry may be about to undergo very large changes. The economics are therefore inchoate. The last word was used by Milton to describe the status of the world before God said 'Let there be light.'

Predictions by several fuel cell organisations for incomplete systems are not in unison, but all see economic improvements coming from mass production, notably ITM. That will be very necessary to meet the intense competition as improved vehicles with new engine schemes enter the market. The industry needs complete fuel cells to achieve competitive performance and any kind of mature economics. The difficulty of the situation is highlighted by the fuel cell bus which saves local pollution on the road, but generates at the power plant stack more pollution than its saves locally. The economics of that unprofitable, untenable situation do not bear examination. Fuel cell industry investors are large and wealthy, a status they are going to need. The complete fuel cell and the hydrogen mine are, in the author's opinion, major development problems to be tackled.

Appendix A

Equilibrium Thermodynamics of Perfect Fuel Cells

The previous chapters led to a situation in which the detailed calculation of the fuel chemical exergy of CO and H_2 had to be carried out as a function of pressure and temperature, following and expanding upon Barclay (2002) and using the equilibrium diagram thereof. During those calculations the work of the University of Pennsylvania and North Western University on direct methane oxidation obtruded. A calculation of the chemical exergy of methane became of interest. There were some difficulties and approximations, and these led to the author doing the calculation by two novel and independent routes, which gave approximately the same answer. A reversible method of consuming methane can of course be reversed, and one of the two routes was an apparent base for making power, or power and hydrogen, a solution to the hydrogen mine problem. Moreover, as a power production route, the performance was of very high efficiency, and threatened the very existence of fire and the heat engine as a route to power production.

A concurrent development has been the proposed use of fuel cells integrated with gas turbines, as a power production route. The use of complete fuel cells in such a system would lead to high efficiency, and accordingly the system is reviewed.

A last-minute incursion on the scene has been the announcement by Harvard University of its initial attempt to develop a low-temperature coal fuel cell (Weibel *et al.*, 2005a, b). The language of the two papers produced is that of isothermal oxidation. There is resemblance to the

Fuel Cells, Engines and Hydrogen – An Exergy Approach   Frederick J. Barclay
© 2006 John Wiley & Sons, Ltd

anolyte/catholyte terminology of Regenesys, but the performance analysis depends on calculated combustion thermodynamics. Moreover, the flow sheet given by Harvard, Figure A.8 below, does not include an equilibrium diagram, or lead to economic choices on how far to far pursue perfection. The author is in correspondence with Harvard about this.

A.1 THERMODYNAMIC PREAMBLE TO THE FUEL CELL EQUILIBRIUM DIAGRAM

Where the bee sucks, there suck I. In a cowslip bell I lie.
There I'm couch'd, when owls do cry. On a bat's back do I fly.
                                                        William Shakespeare

A.1.1 Equilibrium Thermodynamics

Shakespeare's fairyland is mirrored in equilibrium thermodynamics: all is simplicity and perfection! For fuel cells, the gist of such a theory, tackled by Gardiner (1996), but challenged by Appleby (1994), is that the irreversible losses inherent in practical systems must be separated and evaluated. Then a comparison of practical with perfect, via a summation of the losses, leads to a calculated and understandable efficiency. The latter is an underlay to the economics, the final arbiter. The notion that the calorific value of the fuel, as distinct from its much larger chemical exergy, is a basis for performance calculations has been dismissed by Barclay (2002). In the foreword of this book, it is predicted that the novel ideas herein will get over, but rather slowly. But the ideas are not challenged.

This appendix uses frequently the terms 'work potential' and 'exergy' (Barclay, 1998; Kotas, 1995). For both of these, 'The go of things' is a colloquial interpretation. For example, early contemplation of a hydroelectric power project involves, as a work potential, the depth of water behind the proposed dam, or the difference in elevation of the upper and lower reservoirs, Δh. The latter is a convenient approximation, which takes no account of the small increase in water density with depth. In the turbine, at the dam base, the enthalpy of the compressed water is reduced to the atmospheric pressure value. With very low Δt during expansion, it makes little practical difference whether the change is isentropic or isothermal. For accuracy, and for understanding, however, the expansion must be isothermal. Otherwise the turbine discharge will not quite be in temperature equilibrium with the environment. The thermo-dynamic argument is that the maximum, steady flow, power output, is

$$\Delta G = \Delta(H - TS), \quad \text{with } T = \text{constant},$$

derived numerically from tabulated properties of compressed water. ΔG is the universal answer to the problem of determining the work potential of a steady flow, isothermal, physico-chemical process, the unachievable ideal to which a practical near isothermal process, with friction and turbulence losses, aspires (Atkins, 1995).

When it comes to assessing the **maximum possible work potential** of the **ambient** isothermal electrochemical oxidation of hydrogen, it is necessary to envisage the apparatus in which such a reaction could be organised, much more than a fuel cell (i.e. Figure A.1). The figure was drawn by envisaging reversible steps in a process designed to include and envelope isothermal oxidation in a fuel cell. Notional components are used, which are not immediately available. For example, selectively

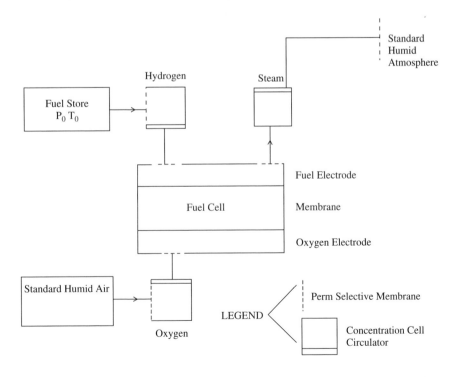

Equilibrium Diagram For Hydrogen Fuel Cell

Figure A.1 Ambient pressure and temperature hydrogen fuel cell

permeable membranes and concentration cells handle each reactant and product. Imperfect examples of these will undoubtedly appear, but those in Figure A.1 are perfect.

It must be realised that the equilibrium of the central fuel cell occurs at one set of concentrations only (Atkins, 1995), which are identical to the concentrations calculable, and also tabulated in Chase *et al.* (1985), for the **dissociation** of the product into the reactants. If an alternative concentration were considered then it would have to live together with the dissociation equilibrium, which would promote perpetual motion, and would be an absurdity. In the JANAF thermochemical tables (Chase *et al.*, 1998) the all-important equilibrium constant is tabulated as a function of temperature and pressure for the relatively difficult isothermal oxidation of hydrogen, involving the fugacity of the imperfect or non-ideal steam/water product. When the apparatus in Figure A.1 operates at ambient temperature it comprises a fuel cell allied to three concentration cells, one each for hydrogen, oxygen and water vapour. There are associated perm-selective membranes. At high temperature and pressure, see Figure A.2 below, three additional isentropic circulators are needed. Membrane technology is almost ready for such equipment (Dyer *et al.*, 2000), so that what has been envisaged, although merely a calculation route, is nearing the edge of practicality.

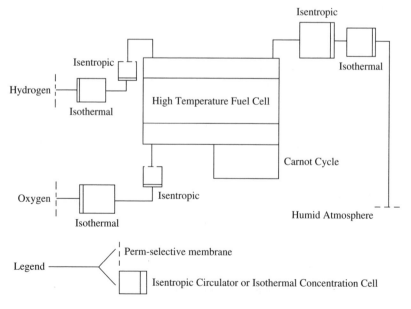

Figure A.2   High-pressure and high-temperature fuel cell

The numerical result is the summation of many ΔG terms. Given a number for $\Sigma \Delta G$, and a name, 'the fuel chemical exergy', the problem opens up to become one of evaluating the irreversible losses inherent in practical systems, and so comparing practical with perfect. There is good education in understanding equilibrium systems, which, in the absence of economics and irreversible thermodynamics, are less complicated than practical systems. Components of equilibrium fuel cells can be endowed with 'magic' properties: for example, no secondary chemical reactions at electrodes, notably involving peroxides and superoxides at oxygen electrodes; zero electrical resistance; ideal perm-selective membranes; no molecular diffusion through the electrolyte; and the like. The author uses 'superbium' and 'idyllium' as convenient shorthand for electrode and electrolyte materials with perfect performance in any given situation. The theory and supporting calculations in this appendix, are, in places, new and original. They are based on exergy analysis (Barclay, 1998; Kotas, 1995), which distinguishes between pressure/temperature difference work potential, ΔB (thermomechanical exergy) and isothermal work potential ΔG (chemical exergy). ΔB is used in the analysis of equilibrium systems with high-temperature fuel cells. Such systems are of greater complexity than the ambient system above and, it turns out, with an equilibrium thermodynamic disadvantage. In practice, they have the advantage of reduced overvoltages and reduced losses in the non-equilibrium power generating regime, already familiar in existing fuel cells.

A.1.2 Steady Flow Equilibrium Fuel Cell Equilibrated via Concentration Cells with the Environment and both Internally and Externally Reversible

At this stage it is sensible to provide a reminder for readers who are chemical engineers and familiar with 'process integration theory' for cyclic processes involving actual irreversible heat exchanger networks, thermomechanical exergy B, and pinch points between rising temperature and falling temperature streams. See Barclay (1998), Chapter 7 and references therein. Linhoff's *et al.* form of analysis as in Barclay (1998) does not in any way relate to non-cyclic equilibrium **isothermal** oxidation using chemical exergy G, as in Figures A.1 and A.2.

Figures A.1 and A.2 need the comment that whilst water is plentiful, hydrogen fuel is unobtainable in nature, being a reactive, diffusive gas. Manufactured hydrogen, by convention, is stored at standard conditions, P_0/T_0. Manufactured carbon monoxide would be stored in the same way. Oxygen from the air is obtained at its standard atmospheric partial pressure PP_0. The fuel and air concentrations are reduced to the rarefied equilibrium conditions, given precisely in the JANAF thermochemical tables. (Chase *et al.*, 1998). Water vapour is delivered to the atmosphere at standard relative humidity, regardless of whether the cell itself produces steam/water or steam. In the event that the cell produces steam/water at standard conditions, the circulators draw off vapour 'above' the water, so that the net exchange of latent heat with the environment is zero. The exergy or Gibb's potential change of the isothermal, latent heat, phase change is also zero,

$$\Delta G = H - h - T(S - s) = 0$$

The central feature of Figures A.1 and A.2, which are networks of reversible processes including isothermal exchanges of energy with the environment at zero Carnot efficiency, is an idealised fuel cell with pure water electrolyte: the simplest calculation path. The heterogeneous and interfacial, equilibrium, electrochemical reaction is at selected pressure and temperature. The chemical reaction does work via bond force reorganisations, which are the agents of transformation from reactants to product, **at equilibrium concentrations**. Use of the latter is newly introduced in this book. The calculation of the maximum work of reversible isothermal electrochemical oxidation cannot otherwise be accomplished. The isothermally enclosed electrochemical reaction is equilibrated with the fuel, oxidant and product via the reactant and product circulators and a Carnot cycle. The equilibrium reaction is the source of **all three kinds of work potential** in the diagram: electrochemical, circulator and Carnot. The electrochemical charge transfer reaction between hydrated ions and the reactants in the electrolyte (Marcus, 1982) generates the electric potential difference between the fuel cell electrodes, and drives electron flow in the external circuit to and from the electrodes. It also creates the **equilibrium** concentration differences relative to the environment, of the reactants and of the product, at the electrodes, and hence the circulator/concentration cell power production or consumption. At the same time it supplies reversible heat (work potential) to the work-producing Carnot cycle. The total pressure within the isothermal enclosure is the sum of the reactant and product fractional pressures as determined by the equilibrium constant, for the chosen operating conditions. The equilibrium constant dictates the parameters of the equilibrium reaction, namely sparse reactants,

and plentiful product. The same equilibrium constant governs dissociation of the product (Chase *et al.*, 1985; 1998). At equilibrium, and
zero reactant consumption, ion migration in the electric field between
the electrodes builds a concentration difference, and hence ion diffusion
such as to balance out migration. Away from equilibrium, there are
no such electrochemical balances. Conditions are unbalanced! There is
current flow in the external circuit, driven by net ion diffusion above
migration.

Figure A.1 shows dissociation in saturated steam, necessarily associated with water. Steam occupies the entire available volume as if the
water were not there. Omitting ice crystal phenomena (Koryta, 1991;
1993; Grahame, 1957), an associated complex of steam/water equilibria
is also shown, for which thermodynamic data are in Chase *et al.* (1998).
Proton exchange (bottom right) occurs when proton tunnelling (fast
process) is allowed between favourably oriented (slow process) water
molecule/hydronium ion pairs (Koryta 1991; 1993). For calculation purposes, at zero-current equilibrium, it is convenient to select pure water
as the electrolyte. Water is marginally self-ionised (Koryta, 1991; 1993),
a more than sufficient electrolyte for zero-current equilibrium. For simplicity of calculation, it is convenient if the electrolyte and the product
are identical. Practicalities such as the introduction of catalysts for vigorous exchange currents, or strong acid or alkali electrolytes for enhanced
power production, can then be handled as perturbations (Bacon, 1969).
When the operating conditions get beyond aqueous electrolytes, and
into the molten carbonate or solid oxide range, the calculations still
relate to hydrogen oxidation, leading to a dissociated steam product. In
no case is any underlying half reaction allowed to be other than single
stage and reversible. Irreversible electrode reactions are quite another
non-equilibrium topic (Ralph and Hogarth, 2002; Ralph *et al.*, 2002;
2003).

In reversible, conservative dissociation, bonds (forces) are disman-
tled and remade on a balanced basis, with zero net work potential. The cell reactions of Figures A.1 and A.2 are within equipment
enabling the performance of work, but the equilibrium constant makes
no distinction between circumstances in which, at equilibrium, all the
available work potential can be garnered, or none can be garnered
(Figure A.1).

In nature's photosynthesis process the **power of sunlight** is used to
force together **isothermally** a flow of carbon dioxide and water to form
hydrocarbons and oxygen. That is the opposite of isothermal power
production in a fuel cell. It is correct to refer here to solar power, but

since the delicate vegetation does not get burnt, solar energy is erroneous usage for solar exergy or power.

For a single reversible process between two sets of fixed conditions, the work is independent of the reversible path. However, in a network of reversible processes, such as Figure A.1, alteration of the pressure and temperature of the isothermal enclosure alters the pressure ratio of, for example, the fuel isothermal expander. The power output of Figure A.2 is therefore variable and not a constant, merely because it is reversible. The maximum power, the fuel chemical exergy, is obtained from an electrochemical reaction at standard temperature, T_0, and sum of reactant and product pressures, P_0, with isothermal expanders only and without a Carnot cycle.

The concentrations within the dissociated vapour in Figure A.1 are numerically determined via equilibrium constant values taken from the JANAF tables (Chase *et al.*, 1998). Because of balanced migration and diffusion of ions, the equilibrium cell is an ion concentration cell: that is, there is an ion concentration difference between the electrodes, pro rata to V_n. For steam/water the JANAF tables use fugacity rather than pressure (Atkins, 1995; Moran and Shapiro, 1993) (to represent the very non-ideal product, minimally dissociated water/steam). Moran and Shapiro (1993) include a generalised fugacity chart for condensable fluids, such as steam/water. Forward fuel cell reactions have a positive $\log_{10}K$. At this stage, it can be repeated that, from experience of such calculations, there is a high vacuum of reactants relative to a substantial concentration of product, for any likely chosen equilibrium operating conditions. In alternative words, the dissociation equilibrium is always to the extreme right.

A.1.3   The Duty of Circulators

The circulators must supply to the isothermal enclosure in Figure A.1 reactants at high vacuum conditions and at the desired temperature T_1. If $T_1 = T_0$, isentropic circulators are absent, and the product isothermal circulators must draw from the high steam concentration associated with the cathode water in the isothermal enclosure, and deliver to the low concentration in the standard atmosphere. Inlet and outlet perm-selective membranes ensure that circulator pairs handle only one substance. There is no change in concentration across the three perm-selective membranes having access to the isothermal enclosure. Nor is there a concentration change at the membranes having access to the fuel or to the atmosphere. In brief, there are no irreversibilities inherent in the diagram. Entropy changes associated with the electrochemical reaction, the Carnot cycle or with the isothermal circulators are reversible entropy changes, denoting heat exchange via zero temperature difference. In the absence of irreversibilities, work potential is at a maximum for the selected T_1. As drawn, with local membranes at the isothermal enclosure boundary, there would be uneven concentrations inside the isothermal enclosure, and an irreversible mixing process. It is beyond the illustration to show the innumerable finely distributed membranes which would remove that irreversibility, something like the huge, tennis-court-sized, multiply subdivided membranes of the human lung.

A.1.4   The Fuel Cell Isothermal Oxidation Reaction

The fuel cell electrochemical reaction has, with hydrogen fuel, a negative entropy change (Chase *et al.*, 1998) corresponding to a heat output. There needs to be a Carnot engine to intercept and use that heat, reversibly, as in Figure A.2. In the event of a positive entropy change (Koryta, 1993), a heat pump is needed. Reversible entropy change of either sign represents the fact that the summation of bond forces in the reactants is not the same as in the product. The difference is in no sense an irreversibility, and the use of $(H - TS)/H$ as an efficiency is incorrect. TS is **not** an irreversibility.

A.1.5   Calorific Value Analysis

It is, up to 2006, customary on a worldwide basis (Barclay, 2002) to use the calorific value or enthalpy of combustion (energy) as an efficiency basis for fuel cells, a number with which to compare the practical output (work). It is seen as convenient to use a common standard with heat engines. Calculated efficiencies around 40–60% are common, but fallacious. Moreover, there is no way in which the calculated work output of the equilibrium Figure A.1 could be measured in a combustion calorimeter. Combustion is a fundamentally irreversible process which destroys large amounts of work potential during the slowing down by multiple collisions of high-velocity product atoms, whereas Figures A.1 and A.2 are drawn to avoid all such destruction, so ensuring optimum performance. This appendix asserts that the rational numerical basis of fuel cell efficiency is the relatively large fuel chemical exergy (work potential). The same quantity is the rational efficiency basis for heat engines, and since the integration of high-temperature fuel cells and gas turbines is a serious future project, the unwelcome prospect must be faced of changing the basis of both fuel cell and engine efficiency. The good news is that the work potential of the fuel cell is much larger than the calorific value (multiplied by the Carnot efficiency), but the fuel cell efficiency is much lower than is usually calculated, indicating that the full potential of the fuel cell is not being realised. Engine efficiency also goes down, and greatly so, major irreversibility having been recognised. Combustion is a chain reaction sequence of energetic collision reactions in a cool mixture, with obvious thermal degradation between hot products and their milieu. The altered perspective of fuel cells versus engines helps greatly with the appreciation of the integration of fuel cells with engines. It becomes clear that the analogy for the turbine/fuel cell combination is that the gas turbine is the mouse under the rich man's table, picking up the crumbs after isothermal oxidation in the fuel cell.

A.1.6 Definition of Nernst Potential Difference, V_n

Knowing the concentrations at equilibrium (see Section A.1.2), it can be deduced that the isothermal circulators of oxygen and hydrogen are expanders, while the associated isentropic machines, since they provide a temperature rise, are compressors. On the product side, both isothermal and isentropic circulators are expanders. Equilibrium is achieved by adjusting the circulators until equilibrium electrode concentrations are reached, and the electric current is zero. The equilibrium cell is then insensitive to being open-circuited, and has the Nernst potential difference V_n (Eqn (4.1)). The circulators (concentration cells) will in fact be at zero flow, since there is no reactant utilisation, and hence no net product. An equilibrium cell isolated from fuel, atmospheric oxygen, and water vapour is at zero flow and at V_n. If some reactant flow is allowed the electrode concentrations change, and the electrode potential difference must reduce from V_n. That change is superposed on any changes due to current flow, and is part of the steep slope of the initial part of the voltage–current characteristic, Figure 6.5.

A.1.7 Summation of ΔG in Figure A.2 for Calculation of Hydrogen Chemical Energy

The details of the isothermal and isentropic circulator work of Figure A.2 are given in the introduction of this book, points 14 and 15, together with the underlying theory of the associated Carnot cycle, as

Temperature T1

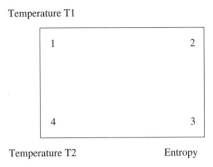

Temperature T2   Entropy

Figure A.3   Details of the Carnot cycle

a way of highlighting immediately the new central topic of reversible equilibrium isothermal oxidation.

The reader should refer to the introduction for these points, which are not repeated here.

The **Carnot cycle is illustrated in Figure A.3.**

As the electrical work of the electrochemical reaction in Figure A.2 declines with increasing temperature, there is a corresponding increase of Carnot work and a change of the substantial circulator power. At standard conditions the isentropic circulators are redundant, as is the Carnot cycle.

A.1.8   Energy Balance Versus Exergy Account

The analysis in the previous section gets at the work done in a reversible network. It is an exergy account. Looking at the network from a first-law, energy (or enthalpy) balance viewpoint, it can be observed, for a perfect gas, that the isothermal circulators are self-balancing, since the same logarithmic expression (see the introduction, point 14) gives power and energy interchange with the environment. The isentropic circulators are presented as an enthalpy balance in the introduction, point 14. The reversible equilibrium fuel cell power can only be derived as ΔG_t by second-law analysis. No additional fuel cell information can be had from an energy balance.

In the case of a practical heat engine an energy balance and a flow balance are made up based on knowledge of turbine blade, heat exchanger and other characteristics. An exergy account, or entropy balance, then establishes the detail of the losses or irreversibilities. No such losses occur in the equilibrium of Figure A.1. See Figure 2.2 of Barclay (1998).

A.1.9   Equilibrium Concentrations

The discussion here follows section 9 of Atkins (1995).

The to and fro, equilibrium homogeneous dissociation reactions, and the heterogeneous Faradaic electrode reaction, which co-exist in an equilibrium fuel cell, have the equation

$$H_2 + \frac{1}{2}O_2 \Leftrightarrow H_2O$$

with one and the same equilibrium constant. The convention is that the rightward reaction is a forward increment away from equilibrium impossible to realise. However, larger pressure ratios may be attainable via concentration cells. The upshot, with practical air-breathing cells, in the absence of circulators, is that there are strong concentration gradients capable of attracting fuel and oxygen to the electrochemical reaction of the fuel cell, via irreversible diffusion.

Note that the muscular motive power of the human body (see the introduction to this book) comes from the **isothermal** hydrolysis of adenosine tri-phosphate to di-phosphate (Atkins, 1995), a wet reaction in which little power is associated with reactant and product handling. Nature has evolved to a position of elegant fuel economy. Nature also has to undertake a complex chemical manufacturing task to provide automated self-maintenance of the body. Chemical reactions are isothermal at blood temperature, along the same lines as the Regenesys system of Chapter 2.

The confidence and experience gained in producing equilibrium diagrams Figures A.1 and A.2 led the author to produce the more complex Figures A.4 and A.4, further below. Any calculation of chemical exergy involves attempt at a new equilibrium diagram.

A.2   UTILISATION OF EQUILIBRIUM DIAGRAM FOR CALCULATION OF CHEMICAL EXERGY

A.2.1   Chemical Exergy of Hydrogen and Carbon Monoxide Derived Using Fugacity-Based Equilibrium Constants

To calculate the equilibrium constant for isothermal oxidation of hydrogen is much more ponderous than for carbon monoxide (Atkins, 1995; Moran and Shapiro, 1993). In the latter case the reactants and the product are near enough ideal gases. But water is a non-ideal, or real, fluid. Accordingly the equilibrium constant must be calculated on the basis of fugacity as in Chase et al. (1998).

Simplistically

$$K = (P_{H_2O}/P_\theta)/(P_{H_2}/P_\theta)(P_{O_2}/P_\theta)^{1/2} \qquad (A.1)$$

In reality the P (pressures) should be F (fugacities), excepting only $P = F$ (arbitrarily) at standard pressure P_0 and temperature T_0. The subsequent calculation involves the equation of state for steam/water. Fortunately, the results of the calculation are available in Chase *et al.* (1998). The tables there are displayed as examples in the introduction to this book. From the data in the tables it can be shown (see below) that the equilibrium constant, Eqn A.1, being independent of the fugacity, is not independent of the pressure. On the other hand, the carbon monoxide equilibrium constant is pressure independent. In fact the Nernst potential difference, V_n, and pro rata the equilibrium constant of the equilibrium hydrogen fuel cell decline with increasing pressure. It is worth emphasising here that equilibrium theory and equilibrium constants relate only to equilibrium chemical reactions, such as a zero-current fuel cell. The occurrence of such obtrusively irreversible processes as chain reaction combustion, or explosions in some hydrogen air mixtures, is not covered. It is expedient to note here that K is a constant. Hence the composition terms P/P_θ in Eqn (A.1) are a function of pressure. See the composition calculations below.

A.2.2   The SHE and the Hydrogen–Oxygen Cell: Mixed Versus Unmixed Potentials

The standard hydrogen electrode (SHE) (Koryta, 1991; 1993; Bard and Falkener, 1980) involves bubbling hydrogen over a platinum black electrode. The platinum black or platinized platinum surface, with extended area, chaotically crystallized surface, multiple dislocations and concentration of adsorbed hydrogen atoms, is bordered by a cloud of tunnelling electrons (Schmickler, 1993), the ensemble confirmed by experiment as an effective catalyst for electrode reactions involving electron transfer (Marcus, 1982). The SHE is a practical, precise, stable, controllable piece of laboratory equipment, the potential difference of which, relative to a 'Bacon' oxygen cathode is very close to 1.23 V. It is suggested in Liebhafsky and Cairns (1968), Chapter 9, that Bacon's semiconductor electrodes, devised as anickel corrosion protection system, had the very fortunate electrocatalytic side effect of preventing formation of peroxides of hydrogen (H_2O_2 and HO_2), or of catalysing their decomposition. That simplified and made reversible the electrode chemistry, simultaneously avoiding mixed potentials. Other oxygen cathodes, for example those employed in current PEFCs, have multistep irreversible chemistry, and a resulting, mixed, open-circuit potential difference, of, say, 1 V (Ralph et al., 2002). The exchange current at such an electrode cannot have a simple reversible chemical equation.

The attainment of 1.23 V in a practical piece of equipment devoid of circulators, exactly as in Table A.1, calculated for Figure A.1, raises the question, 'How can the two, namely Figure A.1 and the experimental arrangement, without circulators/concentration cells produce the same potential difference?' The answer has to be that the laboratory measurement at open circuit requires no fuel or oxygen flow. However, the equilibrium dissociation concentrations of the reactants are always available from the steam product and from the vapour of the aqueous electrolyte. The laboratory arrangement may be described as internally reversible or equilibrated, but externally irreversible, since there is no means, even slightly away from equilibrium, of reversible reactant supply or product evacuation. The missing circulator routes are replaced, away from equilibrium, by irreversible diffusion in a concentration gradient. The previous sentence implies that the fuel cells of Chapters 4–6 are, without circulators, low-efficiency devices. On the other hand, circulators, with associated perm-selective membranes, are not a cheap and easy development.

Fick's law of diffusion, $J = -D\partial C/\partial x$, applies when there is a source and a sink, with a vector current between then proportional to the concentration gradient. An equilibrium electrochemical reaction on open circuit is not a sink for reactants nor a source of product. It is a zero-flow blockage, albeit with balanced exchange currents. The flow of ions in the cell is a balance: the drive V_n is balanced by an opposing concentration difference.

Carbon monoxide and hydrogen develop different potential differences (Table A.1) when oxidised in similar fuel cells, since their

Table A.1   CO and H_2 equilibrium parameters

		$Log_{10}K$	$\Delta G = -RT \ln K$ (kW s mol^{-1})	$V_n = -\Delta G/\nu F$ (V)
CO 1 bar	298 K	45.066	−257.2	1.33
1 bar	1300 K	6.818	−169.7	0.88
10 bar	1300 K	6.818	−169.7	0.88
H_2 1 bar	298 K	41.546	−237.1	1.23
1 bar	1300 K	7.063	−175.8	0.91
10 bar	1300 K	6.063	−151.0	0.78

equilibrium constants are different. It is practical, but irreversible, to consume them together in one high-temperature fuel cell. Hence, the chemical exergy of a mixture of the two gases cannot be calculated using such a mixed process, because an irreversibility is involved in combining the two different V_n potential differences into a 'mixed potential difference', involving local electrode circulating currents. The exergy calculation must use separate cells for each fuel. See Figure A.4 where half reactions have been allocated. The aim is a calculation route, without irreversibilities. Use is therefore made in the figure of mythical gas electrode structures of porous 'superbium', and the equally mythical electrolyte 'idyllium'. With the latter two, ideal equilibrium operation can, without question, be obtained at any operating conditions.

A.2.3   Use of JANAF Tables

A first use of the JANAF tables (Chase *et al.*, 1998) shows, as in Table A.1, equilibrium constants as a function of pressure and temperature, for the fuels CO and H_2. The higher conditions chosen are representative of SOFCs, the highest temperature practical system. Moreover, no interpolation in the JANAF tables is needed. The page numbers in the JANAF tables are CO, p. 628 minus p. 626; H_2, p. 1260; and H_2O, pp. 1275–1276. For oxygen the source is p. 1667.

The bold 1.23 V value has been calculated from the JANAF tables for Figure A.1 ($\nu = 2$, F = 96 484.6) The identical laboratory potential difference is discussed in the previous section. Table A.1 data are for infinitesimally forward reactions. In the fourth column $\nu = 2$ for H_2 and CO (ν is the electron transfer per molecule of fuel). ΔG requires a multiplier of 10^3 for the V_n calculation, arising from the units used in (A.1). The expressions for ΔG and V_n are explained in Atkins (1995).

A.2.4   High-Temperature Fuel Cell and Temperature Changes

Note that the electrical work potential consistently declines with temperature, while the Carnot work potential must increase. Fuel cell electrical work potential and Carnot work potential are added in Gardiner (1996) and termed the 'combined work'. That example is followed here. The combined work has to be calculated separately, and added to the circulator work potential. The total is the fuel chemical exergy, ξ_{ch}.

A.2.5 Effect of Fugacity, Perfect and Imperfect Gases

In Table A.1, the obtrusive difference between carbon monoxide and hydrogen is the zero response of the carbon monoxide $\log_{10} K$ to pressure, in contrast to the significant response of hydrogen, for which $\log_{10} K$, the electrical output ΔG, and V_n each decline with pressure. The same pressure sensitivity differentiates hydrogen fuel in the next section.

A.2.6 Calculation of CO and Hydrogen Oxidation

Using Chase *et al.* (1998) for $(\Delta G, S)$ values, the increase of Carnot cycle work versus the reduction in fuel cell electrical work is now calculated for both fuel electrochemical reactions, at conditions $P_0 T_0$ and above. Included in brackets are $(-\Delta G\,\mathrm{W\,s\,mol^{-1}}, S\,\mathrm{kJ\,mol^{-1}K^{-1}})$:

$$CO + \frac{1}{2}O_2 = CO_2 \text{ and } H_2 + \frac{1}{2}O_2 = H_2O$$

CO 1 bar, 298.15 K: $(-137.2,\ 197.7) + \frac{1}{2}(0,\ 205.1) = (-394.4, 213.7)$
Hence $\Delta G_0 = -257.2 = RT \ln K$ (Table A.1), $\Delta S_{rev} = -86.49$ and $\Delta C_t = 0$.

H$_2$ 1 bar, 298.15K: $(0, 130.7) + \frac{1}{2}(0, 205.1) = (-237.1, 69.95)$
Hence $\Delta G_0 = -237.1 = RT \ln K$ (Table A.1), $\Delta S_{rev} = -163.3$, and $\Delta C_t = 0$.

CO 1 bar and 10 bar, 1300 K: $(-226.5, 243.4) + \frac{1}{2}(0, 252.9) = (-396.2, 283.9)$ (pressure-independent, ideal gas)
Hence $\Delta G_t = -169.7, \Delta S_{rev} = -85.9$ and $\Delta C_t = \eta_c T\Delta S_{rev} = 1002/1000 \times (-85.9) = -86.1$; $\Delta G_t + \Delta C_t = -255.8 \approx \Delta G_0 = -257.2$ (reason for small difference unidentified).

H$_2$ 1 bar, 1300 K: $(0, 174.3) + \frac{1}{2}(0, 252.9) = (-175.8, 244.0)$ (pressure dependent)
Hence $\Delta G_t = -175.8 = RT \ln K$ from Table A.1, $\Delta S_{rev} = -56.8$ and $\Delta C_t = 56.9$; $\Delta G_t + \Delta C_t = 232.7 \approx \Delta G_0 = 237.1$ (reason for small difference unidentified).

H$_2$ 10 bar, 1300 K: $(0, 174.3) + \frac{1}{2}(0, 252.9) = (-150.9, 224.9)$ (pressure dependent)

Hence $\Delta G_{tp} = -150.9 = RT \ln K$ from Table A.1, $\Delta S_{rev} = -75.9$ and $\Delta C_{tp} = -76.1$; $\Delta G_{tp} + \Delta C_{tp} = -150.9 - 76.1 = 227 \neq 237.1 = \Delta G_0$.

In the fugacity-based thermodynamics of hydrogen fuel the match between ΔG_0 and $\Delta G_{tp} + \Delta C_{tp}$ is inexact, due to the influence of pressure on non-ideal water properties.

A.2.7 Note on Use of JANAF Tables

The JANAF thermochemical tables deal with the two forward solid carbon oxidation reactions producing CO and CO_2. The CO oxidation is a matter of subtraction for the equilibrium constant.

A.2.8 Equilibrium Concentrations for Fuel Isothermal Oxidation

The calculations for CO and H_2 are along the lines of Bacon (1969), p. 630, except that a binomial approximation is used for small z, eliminating the usual cut and try calculation.

A.2.9 Hydrogen Oxidation at 298.15 K, 1 Bar

At standard conditions only, pressure $=$ fugacity (Atkins, 1995), and (A.1) applies. The equilibrium mixture is identical to the slightly dissociated water vapour, associated with the water at the cathode of the cell. The equilibrium concentrations are going to be a small amount of reactants, hydrogen and oxygen, relative to a high concentration of H_2O:

$$H_2 + \frac{1}{2} O_2 \Rightarrow zH_2 + (z/2)O_2 + (1-z)H_2O$$

Note that z moles of hydrogen molecules remain, after $(1-z)$ moles of hydrogen go into the product. Likewise $z/2$ moles of oxygen remain after $(1-z)/2$ moles of oxygen go into the product.

Hence, number of moles in mixture (the amount of substance) is

$$n = z + z/2 + (1-z) = (2+z)/2$$

The molar fractions in the mixture are

$$\gamma_{H_2} = 2z/(2+z), \quad \gamma_{O_2} = z/(2+z), \quad \gamma_{H_2O} = 2(1-z)/(2+z) \quad \text{(A.2)}$$

For very small z, $\gamma_{H_2} = z$, $\gamma_{O_2} = z/2$ and $\gamma_{H_2O} \approx 1$.

From the JANAF tables via Table A.1 and for the backward reaction (Moran and Shapiro, 1993)

$$\log_{10} K = -41.546 \text{ and } K = 2.844e^{-42}$$

and for small z, Eqn (A.1) reduces to $K = 1/(z \times (z/1)^{1/2}) \times (P/P_{ref})^{1/2}$, or

$$K = 1.414 z^{2/3} (P/P_{ref})^{1/2} \tag{A.3}$$

Hence $z^{-3/2} = K/1.414 = 2.0113e^{-42}$ and

$$z = \gamma_{H_2} = (2.0113e^{-42})^{2/3} = 1.5432e^{-28}, \quad z/2 = \gamma_{O_2} = 7.716e^{-29}$$
$$\text{and} \quad \gamma_{H_2O} = 1 - z \approx 1$$

Note that γ_{H_2O} must be steam. Water molecules present as bulk water occupy trivial volume.

A.2.10   Carbon Monoxide Oxidation at 1 Bar, 298.15 K

The thermodynamics and compositions of dissociation are calculated here. CO is not 'out of court', in an equilibrium, pure water electrolyte, cell with no catalyst. The neutral water can be made slightly acid, hydronium ions, or slightly alkaline, hydroxyl ions. The hydroxyl alternative, Figure A.1, is selected, the opposite to Section A.2.9.

For CO,

$$\log_{10} K = -45.066, \quad K = 8.590e^{-46} \text{ and } K/1.414 = 6.075e^{-46}$$
$$z = \gamma_{CO} = (6.075e^{-46})^{2/3} = 6.928e^{-31}, \quad z/2 = \gamma_{O_2} = 3.464e^{-31} \quad \text{and}$$
$$1 - z = \gamma_{CO_2} \approx 1$$

A.2.11   Hydrogen Oxidation at 1 Bar, 1300 K

For 1300 K a switch to O^{2-} ions is involved. Superheated steam is produced at the anode. The O^{2-} ion concentration gradient climbs from the oxygen cathode to hydrogen anode, to achieve migration/diffusion balance. Thus

$$\log_{10} K = -7.063, \quad K = 8.65e^{-08} \text{ and } K/1.414 = 6.12e^{-08}$$
$$z = \gamma_{H_2} = (6.012e^{-08})^{2/3} = 6.68e^{-05}, \quad z/2 = \gamma_{O_2} = 3.34e^{-05} \quad \text{and}$$
$$1 - z = \gamma_{H_2O} \approx 1$$

A.2.12   Carbon Monoxide Oxidation at 1 Bar, 1300 K

Here ´

$$\log_{10}K = -6.818, \quad K = 1.52e^{-07} \text{ and } \quad K/1.414 = 1.075e^{-07}$$

$$z = \gamma_{CO} = (1.075e^{-07})^{2/3} = 2.14e^{-05}, \quad z/2 = \gamma_{O_2} = 1.07e^{-05} \quad \text{and}$$

$$1 - z = \gamma_{CO_2} \approx 1$$

A.2.13   Carbon Monoxide Oxidation at 10 Bar, 1300 K

Here

$$\log_{10}K = -6.818, \quad K = 1.52e^{-07} \times (P/P_{ref})^{1/2} = 1.52e^{-07}10^{1/2}$$

$$= 4.81e^{-07} \text{ and } \quad K/1.414 = 3.40e^{-07}$$

$$z = \gamma_{CO} = (3.4e^{-07})^{2/3} = 4.85e^{-05}, z/2 = 2.43e^{-05} \text{ and}$$

$$1 - z = \gamma_{CO_2} \approx 1 (\text{anode})$$

The Le Chatelier prediction of pressure favouring the product is quantified. Relative to Section A.2.12, z is reduced.

A.2.14   Hydrogen Oxidation at 10 Bar, 1300 K

Here $\log_{10}K = -6.063$ and $K = 8.65e^{-07}$. Then, as in the previous section, $8.65e^{-07} = 1.414 \times 10^{1/2}z^{1/2}, z^{1/2} = 1.935e^{-07}$ and $z = 3.33e^{-05}$. Thus, $z = \gamma_{H_2} = 3.33e^{-05}, \gamma_{O_2} = 1.66e^{-05}$ and $\gamma_{H_2O} \approx 1$, cf. Section A.2.11. The steam is superheated and there are no water molecules. Pressure favours the product. (The fugacity coefficient is 0.99 (Atkins, 1995), and so $10^{1/2}$ is a valid approximation although derived from ideal gas theory).

Remark

The 1 bar and 10 bar, 1300 K equilibrium concentrations = degree of product dissociation at 1300 K (Chase *et al.*, 1998) are much higher than those at 298 K. The hydrogen reaction at 1300 K is once again pressure sensitive, due to a reduced equilibrium coefficient. The γ values are electrode concentrations at equilibrium, and at the same time represent the dissociated product equilibrium. These γ values lead, in due course, to circulator powers, which are greatly reduced at high temperature and pressure, a reason for selecting high temperature and pressure in practice.

A.2.15 Circulator Power and Chemical Exergy Calculated at Standard Conditions

The three isothermal circulators in Figure A.1 are clearly all expanders, producing power. The isentropic circulators and the Carnot cycle are redundant, having zero temperature difference. The circulator power added to the standard conditions ΔG of Table A.2 will give the **chemical exergy** of the two fuels, CO and H_2.

A.2.16 Fuel Chemical Exergy of CO

The ideal gas calculation for **CO** is performed first (see Section A.2.11).

An isothermal expander has an output of $-RT \ln (P_i/P_f) = \Delta G \, W\, s\, mol^{-1}$.

$$\text{For CO} \quad \Delta G = (8.315 \times 298.15 \times \ln (1/6.928e^{-31}))/1000$$

$$= (8.315 \times 298.15 \times 69.442)/1000 = 172.2$$

$$\text{For } O_2, \quad \Delta G = (1/2)(8.315 \times 298 \times \ln (0.204/3.464e^{-31})/1000$$

$$= (1/2)(8.315 \times 298.15 \times 68.55)/1000 = 85.0$$

$$\text{For } CO_2, \quad \Delta G = (8.315 \times 298.15 \times \ln 1/0.0003)/1000$$

$$= (8.315 \times 298.15 \times 8.112)/1000 = 20.1$$

Hence

$$\textbf{Chemical Exergy of CO} = \Delta\xi_{ch} = 257.2 + 172.2 + 85.0 + 20.1$$

$$= 257.2 + 277.3 = 534.5 \, kW\, s\, mol^{-1}$$

The circulator work potential 277.3 exceeds that of the cell 257.2!

Note that the enthalpy of combustion is $283.0 \, kJ\, mol^{-1}$, an unrelated fact. The product of a Carnot efficiency of, say, 50% and the combustion enthalpy is $143.5 \, kW\, s\, mol^{-1}$ (cf 534.5). The potential superiority of the fuel cell is evident. Referring the efficiency of heat engines to the fuel

Table A.2   Isothermal circulator power, CO fuel, standard conditions

Gas	P_f (bar)	P_i (bar)	Relative flow	Work potential (KW s mol^{-1})
CO	6.928 e^{-31}	1.0	1.0	172.2
$\frac{1}{2}O_2$	3.464 e^{-31}	0.204	0.5	85.0
CO_2	0.0003	≈ 1.0	1.0	20.1

chemical exergy will be difficult to sell, but no sensible efficiency calculations for integrated fuel cell/heat engine systems are possible without such referral.

As discussed above, the pressure ratios of the circulators are impracticably large for mechanical turbo-machines, and accordingly the full chemical exergy of an equilibrium fuel cell would be very difficult to realise, by a factor of more than 2. In the future, however, isothermal concentration cells, generating electrical power from a pressure/concentration difference, might make non-equilibrium circulator power relatively realistic. Suitable membrane technology is becoming available (Dyer *et al.*, 2000). In addition, air-breathing cell technology cuts out any notion of achieving or approaching thermodynamic equilibrium concentrations, and sacrifices the circulator power. The position would be different if the reactants and products were liquids, as in Chapter 2 of this book. Practical gaseous fuel cell designers must accept the facts and concentrate on getting the best bipolar flow plate design.

A.2.17 Fuel Chemical Exergy of Hydrogen

In the corresponding calculation for H_2 an alternative way has to be found to calculate the output of the steam circulator, since steam is highly imperfect. The steam is expanded isothermally at 25 °C, from standard pressure, 1 bar, saturated to standard partial pressure 0.0088 bar. A minor extrapolation of the Mollier chart gives $G_i = H - TS = 2590 - 2555.14$ and $G_f = 2590 - 2745.96$. Hence $\Delta G = -190.82\,\mathrm{kW\,s\,kg^{-1}}$ and $\Delta G = 3.435\,\mathrm{kW\,s\,mol^{-1}}$ (steam circulator).

For the H_2 full-flow circulator,

$$\Delta G = (1/1000)(8.315 \times 298.15 \times \ln{(1/1.543e^{-28})})$$

$$= (1/1000)(8.315 \times 298.15 \times 64.039)$$

$$= 158.8\,\mathrm{kW\,s\,mol^{-1}}$$

For the O_2 half-flow circulator,

$$\Delta G = (1/2000)(8.315 \times 298.15 \times \ln{(0.203/7.716e^{-29})})$$

$$= (1/2000)(8.315 \times 298.15 \times \ln{(1.044e^{29})})$$

$$= (1/2000)(8.315 \times 298.15 \times 63.137)$$

$$= 78.3\,\mathrm{kW\,s\,mol^{-1}}$$

Hence

$$\text{Chemical exergy of } H_2 = \Delta\xi_{ch} = 237.1 + 158.8 + 78.3 + 3.4$$
$$= 477.6\,\text{kW s mol}^{-1}$$

The enthalpy of combustion is $241.830\,\text{kJ mol}^{-1}$, again a smaller quantity, with different units to reflect the difference between power (work rate) and heat (energy flow). Note that different values for the chemical exergy of fuel are given in various texts. Notably there is in Kotas (1995) an alleged equality at Figure 2.12, p. 46, Fuel Chemical Exergy $\approx$ Net Calorific Value. In that figure the van't Hoff equilibrium box contains non-equilibrium concentrations. Every one cannot be right, and the author is sticking to his principles (Barclay, 2002). As an indicator of first-law engine performance, the combustion enthalpy, reduced by the boiler efficiency, has to be multiplied by the cycle efficiency, perhaps 50%. The chemical exergy is work, not Carnot limited energy.

With hydrogen fuel, the large circulator (concentration cell) pressure ratios and unavailable membranes are again obstacles to the realisation in practice of the theoretical work potential.

A.2.18   Real Electrolytes

The addition of, for example KOH, to the water for a strong electrolyte makes a trivial difference to the chemical exergy calculation for hydrogen. The boiling point of the solution is elevated and the steam product expands from a superheat point a little to the left of the saturation line, along the isotherm, making a small difference to the figure of 2.828 in Table A.3.

Table A.3   Isothermal Circulator Power - Hydrogen Fuel- P_0T_0

Gas	P_i(bar)	P_f (bar)	Relative flow	Work potential kW s mol^{-1}
H_2	1	$1.543e^{-28}$	1	158.68
$\frac{1}{2}O_2$	0.203	$7.716e^{-29}$	0.5	78.23
H_2O	$G_i = 2590 - 2555.14$ $= -34.86$	$G_f = 2590 - 2745.96$ $= -155.96$ $\Delta G = -190.72\,\text{kW s kg}^{-1}$ $= -3.43\,\text{kW s mol}^{-1}$	1	2.828

A.2.19 Pure Water Electrolyte

Pure water would, of course, be an impractical fuel cell electrolyte, but at the equilibrium point there is zero current, and a shortage of ions in minimally ionised pure water makes no difference to the equilibrium parameters. For a current generating fuel cell, vigorous ion migration is a requirement. Hence, for example, the KOH in Bacon's fuel cell (Bacon, 1969; Adams *et al.*, 1963).

A.2.20 Hot Pressurised Equilibria

In order to tackle the pressurised, high-temperature, equilibrium systems, the ΔS and the ΔG must be taken from the JANAF thermochemical tables (Chase *et al.*, 1986; 1998). The ΔS determines the reversible heat input to the Carnot cycle. Pressure affects the equilibrium constant for hydrogen, but not for carbon monoxide. Pressure moves the equilibrium concentrations in the direction of the product, affecting circulator power.

A.2.21 Carbon Monoxide Oxidation at 1 Bar, 1300 K

See Section A.2.13 for γ values and Table A.1 for equilibrium parameters from Chase *et al.* (1986; 1998).

For the fuel cell reaction,

$$0 = -CO - \frac{1}{2} O_2 + CO_2$$

$$\Delta_r G = -396.177 + 226.509 = -169.67$$

$$\Delta S = -243.43 - \frac{1}{2}(252.88) + 283.93 = -85.94$$

Hence Carnot cycle output is

$$((T - T_0)/T) \times T\Delta S = (T - T_0)\Delta S = -1002 \times 85.94/1000 = -86.11$$

and

$$\text{Electrical output} = -169.67$$

A.2.22 Circulator Power at High Temperature

The circulator power now has isentropic and isothermal components.

For the isentropic machines,

$$P_1 V_1/T_1 = P_2 V_2/T_2 \quad \text{and} \quad P_1 V_1^{\gamma} = P_2 V_2^{\gamma}$$

Hence

$$(V_1/V_2)^\gamma = (P_2/P_1) \text{ and } V_1/V_2 = (P_2/P_1)^{-\gamma} = (P_1/P_2)^{-1/\gamma}$$

and

$$T_1/T_2 = (P_1/P_2)(P_1/P_2)^{-1/\gamma} = (P_1/P_2)^{(\gamma-1)/\gamma} : P_1/P_2$$
$$= (T_1/T_2)^{\gamma/(\gamma-1)} \text{ while } P_2/P_1 = (T_2/T_1)^{\gamma/(\gamma-1)}$$

Thus

CO isentropic compressor, $P_2/P_1 = (1300/298)^{1.4/0.4} = 173.4$

O_2 isentropic compressor, $P_2/P_1 = 173.4$

CO_2 isentropic expander, $P_1/P_2 = (1300/298)^{1.31/0.31} = 505.1$

A.2.23 Isentropic Circulators

The enthalpy changes for the three isentropic machines are conveniently tabulated in Chase *et al.*, 1998, column 5 of the JANAF tables. The perfect gas enthalpy is a function of temperature alone, and the temperature range is 1300 to 298.15 K:

$$O_2 \text{ compressor } 33.34/2 = 16.67$$
$$CO \text{ compressor} = 31.87$$
$$CO_2 \text{ expander} = -50.15$$

Hence

$$\text{Net isentropic output} = -1.61 \, \text{kW s mol}^{-1}$$

A.2.24 Isothermal Circulators

The expressions for pressure ratio derived immediately above enable calculation of the pressures at the output of the O_2 and CO isothermal machines, and the input of the CO_2 isothermal machine, and hence their individual powers (kW s mol^{-1}):

CO isothermal circulator, $P_f = (173.4)^{-1} \times 3.587 \times 10^{-5}$
$= 2.069 \times 10^{-7}$ (see Section A.2.13)

$$\Delta G = -RT \ln (P_i/P_f) = (1/1000)(8.315 \times 298 \times \ln \, 2.069 \times 10^7)$$
$$= -41.74$$

O_2 isothermal circulator, $P_f = (173.4)^{-1} \times 1.794 \times 10^{-5}$
$= 1.034 \times 10^{-7}$

$\Delta G = -(1/1000)(8.315 \times 298 \times \ln (0.203/(1.034 \times 10^{-7}))) \times 1/2 =$
$-34.8/2 = -17.4$

CO_2 isothermal circulator, $P_i = 1/505.1 = 1.98 \times 10^{-3}$ and $P_f =$
0.0003

$$\Delta G = -(1/1000)(8.315 \times 298 \times \ln 6.6) = 4.68$$

The total circulator power is $-63.82\,kW\,s\,mol^{-1}$, which is less than at standard conditions because of greater dissociation.

A.2.25 Comparative Data for Standard and High Conditions

In the table, the symbol ΔC_t is the output of the Carnot cycle. At $P_0 T_0$ there is no Carnot cycle and its output is 0^* as in the third column. At high temperature 1300 K, there must be a Carnot cycle of significant output (fifth and seventh columns, values marked $*$), different for the two reactions, perfect CO and imperfect pressure-sensitive H_2O.

The example in the next section demonstrates that the full fuel chemical exergy is obtainable at standard conditions only. The equivalent calculations for hydrogen, and for both fuels at 10 bar, are left as exercises for the reader. Without the 'unachievable circulator power', the combined electrical + Carnot powers are fairly evenly matched (Table A.4).

Table **A.4**   These data are given in Table A.4.

Gas or liquid	$P_0 T_0 S$	$P_0 T_0\,\Delta_f G$	$P_0 1300\,K\,S$	$P_0 1300\,K$ $\Delta_f G$	10 bar, 1300 K S	10 bar, 1300 K $\Delta_f G$
$-CO$	-197.7	$+137.16$	-243.43	$+226.51$	-243.43	$+226.51$
$-\frac{1}{2}O_2$	-102.6	0	-126.439	0	-126.44	0
$+CO_2$	$+213.8$	-394.4	$+283.93$	-396.24	$+283.93$	-396.24
$-H_2$	-130.7	0	-174.288	0	-174.29	0
$-\frac{1}{2}O_2$	-102.6	0	-126.439	0	-126.44	0
$+H_2O$	$+69.95$	-237.1	$+244.03$	-175.78	$+224.86$	-150.89
	2-phase	2-phase	Vapour	Vapour	Vapour	Vapour
$\Delta S_{CO}/\Delta C_t$	-86.49	0^*	-85.94	-86.11^*	-85.94	-86.11^*
ΔG_{CO}		-257.2		-169.7		-169.7
$\Delta G_{CO} + \Delta C_t$		-257.2		-255.8		-255.8
$\Delta S_{H_2}/\Delta C_t$	-163.3	0^*	-56.70	-56.81^*	-75.87	-76.02^*
ΔG_{H_2}		-237.1		-175.8		-150.9
$\Delta G_{H_2} + \Delta C_t$		-237.1		-232.6		-226.9

A.3 CHEMICAL EXERGY OF METHANE AND RELATED HIGH-EFFICIENCY HYDROGEN PRODUCTION

A.3.1 Chemical Exergy of Methane

This is explored by two alternative routes, Section A.3.2 and A.3.10.

A.3.2 Route 1: The Manufacture by Faradaic Reform at Equilibrium of Carbon Monoxide and Hydrogen, from Methane, followed by Oxidation in Two Separate Fuel Cells

The skills developed to produce the equilibrium diagram Figure A.1, are now applied anew. Neither hydrogen nor carbon monoxide occur as free substances in nature, where they are immediately oxidized. They must be made and stored, at thermodynamic and economic cost. The reversible thermodynamics are assessed below, using as the basis of calculation a notional, electrochemical, equilibrium, steam reformer, Figure A.4, for comparison with the alternative practical and irreversible combustion-driven reformers.

In principle a chemical reaction can be supplied with exergy by a variety of means. In photosynthesis the required **power** is supplied by the sun's rays. In Figure A.4 electrical power is used in an unfamiliar way, without practical precedent.

The methane reform overall electrochemical reaction, to be electrically driven, Figure A.4, is

$$CH_4 + H_2O = CO + 3H_2$$

When the circulators are slowed down towards equilibrium, the current goes to zero and the reformer is insensitive to being open-circuited. The author felt that the resemblance of this calculation route to the existing practical irreversible steam reformer, plus fuel cell, made it an unavoidable choice, albeit complicated. Subsequently however, other work (Park *et al.*, 1999; Gorte *et al.*, 2000) has appeared, recording experimental fuel cell direct methane oxidation, at useful rates – a new achievement. A second calculation route was born. (Section A.3.11).

For a Faradaic process to operate, there must be half reactions at the anode and cathode, and these are proposed for the reformer calculation

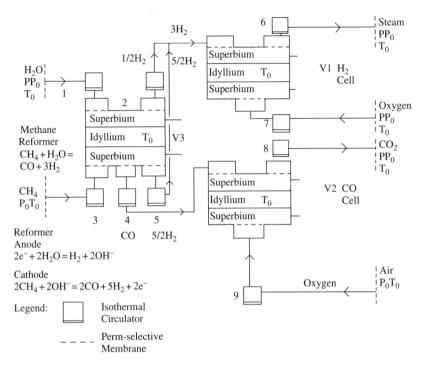

Figure A.4 Combined equilibrium methane reformer and fuel cells for hydrogen and carbon monoxide at standard conditions, P_0 and T_0

route in Figure A.4. For a driven reaction, the electron flow in the external circuit must be in the opposite direction to power production as in Figure A.1. The term for such a system is endergonic or exergy absorbing, by analogy with the better known endothermic, or heat absorbing.

The next step is to inspect the thermodynamic data (Table A.5) for the overall reaction and this is done at three temperatures, 298.15 K, 900 K and 1300 K, representing SPFC, MCFC and SOFC, respectively. Only at $P_0 T_0$ does the thermodynamic data favour the electrically driven reformer. Figure A.4 is for the major calculation route at $P_0 T_0$. The production of hydrogen at both reformer electrodes in the figure is unusual, if the mind has been concentrating on fuel cells. Moreover, $\nu = 1$.

Table A.5　Faradaic equilibrium reform, thermodynamic data

$-CH_4$ ($\Delta G_f/S$)	$-H_2O$ ($\Delta G_f/S$)	$+CO$ ($\Delta G_f/S$)	$+3H_2$ ($\Delta G_f/S$)	ΔG_r	ΔS_r	Temp. (K)
50.77/ − 186.3	237.1/ − 70.00	−137.2/197.7	0/130.7	150.7	333.5	298
−8.6/ − 240.2	198.1/ − 228.4	−191.4/231.1	0/163.1	−1.9	251.8	900
−52.6/ − 267.7	175.8/ − 244.0	−226.5/243.4	0/174.3	−103.3	254.6	1300

Since water is a reactant, fugacity is relevant, and the JANAF tables (Chase *et al.*, 1998) should, for accuracy, be used at other than standard conditions. Note that the reform reaction is not tabulated in the JANAF tables.

In the fourth column of the table the data were multiplied by 3 to get the summations for the fifth and sixth columns. S is given in units of J etc., and G in units of kJ etc. The reformer reaction of the table produces three molecules of hydrogen to fuel the accompanying hydrogen fuel cell, and one CO molecule to fuel the CO cell.

From Table A.4 fuel cell performance at standard conditions is $\Delta G_{H_2} = -237.1$, $\Delta G_{CO} = -257.2$; and from Table A.5 for the P_0T_0 faradaic reformer $\Delta G_r = 150.7$. The net fuel cell work potential of the diagram in Figure A.4 is $-237.1 \times 3 - 257.2 + 150.7 = -817.8$. The cell work potential handsomely exceeds the input of the reformer. At standard conditions, energy exchanges with the environment have zero exergy content, and do not enter the exergy balance. Electrical arrangements are not proposed in Figure A.4, but would take account of the differing potential differences of the reformer and of the two fuel cells. The power of the circulators, Δ_{cr}, has to be added to get at the total power available, $\Delta\xi_{ch}$ (Section A.3.8 below).

At 1 bar and 1300 K, $\Delta G_{H_2} = -175.8$ and $\Delta_{ct} = -56.81$, and for the Faradaic reformer $\Delta G_r = -103.3/3 = -34.43$ and $\Delta_{ct} = 254.6/3 \times (1002/1000) = 85.04$. The Faradaic reformer is now somewhat awkward. It needs a large supply of reversible heat at 1300 K from a heat pump or from the fuel cell, and provides an electrical **output**. Current from the fuel cell would have two functions. Firstly to generate shaft power to **supply** the missing part of the Carnot input of the reformer, $85.04 - 56.81$, and secondly to power a motor generator, with the correct potential difference, to back off the current output of the reformer. Circulator power is not required to complete the transaction.

The intermediate condition of the MCFC at 900 K is left as an exercise for the reader since it is qualitatively the same as the 1300 K case.

A.3.3 Faradaic Reformer, Equilibrium Constant and Equilibrium Mixture at P_0T_0

Since the reform reaction is not tabulated in the JANAF tables, ideal gas theory, an approximation, is the only alternative.

For standard conditions, from Table A.5, $\Delta G_0 = 150.7 = -RT \ln K$. Hence

$$\log_{10} K = -(150.7/1000)/(8.317 \times 298.15 \times 2.303) = -2.639 e^{-5}$$

$K = 0.999 = 1.0$

The mass conservation equation is

$$1CH_4 + 1H_2O = zCH_4 + zH_2O + (1-z)CO + 3(1-z)H_2$$

The total number of moles in the mixture is

$$N = z + z + (1-z) + 3(1-z) = 4 - 2z$$

and the molar analysis is

$$\gamma_{CH_4} = z/(4-2z) = \gamma_{H_2O}, \text{ with } \gamma_{CO} = (1-z)/(4-2z) \text{ and } \gamma_{H_2}$$
$$= 3(1-z)/(4-2z)$$

The equilibrium constant takes the form (see Eqn 14.32, page 627, of Atkins, 19)

$$K = [(P/P_{ref}/(4-2z)]^{1+3-1-1} \times [(1-z) \times (3-3z)^3/z^2]$$

with $P/P_{ref} = F/F_{ref} = 1$ at standard conditions. Hence

$$K = 27(1-z)^4/4(2-z)^2 = 6.75(1-z)^4/(2-z)^2 = 1$$

and by cut and try, $z = 0.155$ and

$$\gamma_{CH_4} = \gamma_{H_{20}} = 0.042, \ \gamma_{CO} = 0.229 \text{ and } \gamma_{H_2} = 0.687$$

with the hydrogen flow split 1 to 5 between anode and cathode.

A difficulty in the above ideal gas analysis is apparent because the H_2O has two-phases. The steam pressure in the fuel cell mixture must in fact be the saturation pressure corresponding to 298.15 K, namely 0.03 bar, rather than 0.042 bar. The missing H_2O is present as water molecules of minimal volume. The whole calculation is approximate, and the correction cannot be guessed. But by experience, the calculation is insensitive to small changes, such as 0.03 versus 0.042 above.

The equilibrium is somewhat product oriented at standard conditions, but would be less so at higher temperature and at pressure.

A.3.4   Electrical Reformer

In the combination of equilibrium Faradaic reformer and fuel cells at standard conditions, the reactions are separated by means of circulators, and do not influence each other, except, of course, that an electrical connection from the fuel cells drives the reformer.

A.3.5   Contrast with Practical Fuel Cell Reformer Pairs

In the real non-equilibrium conditions of a present-day MCFC with very successful electrode reform, the cell electrode reaction, voracious for fuel, consumes the reformer product and favourably influences the reform process. The latter turns out to operate well at 600 °C, compared with about 800 °C in a fired reformer coupled, say, to much less voracious hydrogen separation and storage. In the practical SOFC, 1000 °C at the anode promotes excessively vigorous electrode reform, which leads to a local electrode cold spot. There are also stability considerations (Gardiner, 1996). Hence the contemporary movement towards lower SOFC temperatures, via new ceria electrolytes, and interconnect change from ceramic to steel. A PEFC near T_0, must have a combustion-operated 800 °C reformer, since a T_0 electrochemical reform process does not exist in practice.

A.3.6   Efficiency of the PEFC

The equilibrium reformer/cell combination, based on methane and calculated above, is the beginnings of an efficiency basis for the PEFC. An alternative to separate fuel cells for CO and H_2 would have been a Faradaic shift reactor for the CO, eliminating the CO fuel cell. The reader is left to synthesise that route. However, no change in power output would result from such a rearrangement. (Supporting calculations are not shown, but since the reversible paths of the network all terminate at $P_0 T_0$, it must be so.) The reforming of natural gas would be a much bigger calculation, with a reformer for each reformable hydrocarbon supplying the fuel cells for H_2 and CO. Appropriate electrical arrangements would be needed to match the various potential differences. Enhanced circulator arrangements would also be needed. The problem is appropriate for a computer, just as natural gas combustion enthalpies are computer calculated these days.

A.3.7   Practical Faradaic Reformer

Much more would have to be done in the laboratory to investigate the possibility of a practical Faradaic reformer: choice of electrode and electrolyte; the possibility of irreversible electrode reactions; the need for an electrocatalyst. It can be concluded safely that a basis for fuel chemical exergy efficiency calculations exists, namely the Faradaic reformer, fuel cell combination at standard conditions. The reduced performance of the reformer fuel cell combination, at temperature and pressure, can be left as a major exercise for the reader by adding isentropic circulators and a Carnot cycle to Figure A.2.

A.3.8   Circulator Power

Table A.6   The circulator power for Figure A.4 is given in Table A.6.

Isothermal circulator number	P_i (bar)	Relative flow	P_f (bar)	ΔG (W s mol^{-1})
1 Steam	0.0088	1	0.030	+3.4
2 Hydrogen	0.695	$\frac{1}{2}$	$1.543e^{-28}$	−78.9
3 Methane	1.0	1	0.042	−7.9
4 CO	0.232	1	$6.928e^{-31}$	−168.5
5 Hydrogen	0.695	$\frac{5}{2}$	$1.543e^{-28}$	−394.6
6 Steam	1.0	3	0.0088	−10.2
7 Oxygen	0.204	$\frac{3}{2}$	$7.716e^{-29}$	−234.8
8 CO_2	1.0	1	0.000294	−20.2
9 Oxygen	0.204	$\frac{1}{2}$	$3.464e^{-31}$	−78.3
Total ΔC_r				−990

$\Delta G = -RT \ln (P_i/P_f) = -8.315 \times 298.15 \ln P_i/P_f$.

A.3.9   Methane Chemical Exergy by Reformer Route

Here $\sum \Delta C_r = -990$ (Table A.6), ΔG reformer $= +150.7$, CO chemical exergy $= -257.2$ and hydrogen chemical exergy $= -237.1 \times 3 = -711.3$. Then summing,

$$\sum(-990 - 257.2 - 711.3 + 150.7) = -1807.8$$

Hence   Methane   chemical   exergy $= \xi_{ch} = -990 - 817.8 = -1807.8 \, \mathrm{W \, s \, mol}^{-1}$

By the direct oxidation route below, the result is $-889.5 - 817.9 = -1707.4 \, \mathrm{W \, s \, mol}^{-1}$. Both results are approximations, the errors being in concentrations and circulator powers. The direct oxidation route error, due to two-phase water production (See Section A.3.11), is larger than in the reformer route, so that the reformer route answer $-1807.8 \, \mathrm{W \, s \, mol}^{-1}$ is the more accurate.

The steam/water enthalpy and entropy values for calculation of ΔG came from the steam tables and Mollier chart. See also Kotas (1995), p. 239, for moist standard atmosphere analysis.

Remarks

1. In other texts, the fuel chemical exergy is thought of as a value independent of temperature and pressure, like combustion enthalpy. Instead it has, above, a maximum at $P_0 T_0$. The major difference in calculation routes is that the author uses equilibrium conditions dictated by the equilibrium constant within the isothermal enclosure of the fuel cell, or Faradaic reformer, whereas other writers put reactants in, and take products out, at standard conditions.

2. The overall equation for the oxidation of methane is

$$CH_4 + 2O_2 = CO_2 + 2H_2O \qquad (A.4)$$

For the process of Figure A.4 the equations are

$$CH_4 + H_2O = CO + 3H_2$$
$$CO + \frac{1}{2}O_2 = CO_2$$
$$3H_2 + \frac{1}{2}O_2 = 3H_2O$$

and totalling:

$$CH_4 + 2O_2 + H_2O = 3H_2O + CO_2$$

or:

$$CH_4 + 2O_2 = CO_2 + 2H_2O$$

which repeats Eqn (A.4).

3. Hence, the two routes, with and without reformer, have the same overall equilibrium thermodynamics at standard conditions, the same electrical power, zero Carnot power, and the same net circulator power. However, both the equilibrium constant calculation routes are approximate. The concentrations are in error as a result. See Section A.3.9.

A.3.10   Route 2: Fuel Chemical Exergy by Direct Methane Oxidation

When this appendix was in preparation, methane electrochemical oxidation had not been achieved. A reformer was essential! That fact influenced the author's choice of initial calculation route for the methane chemical exergy, to be via oxidation via an equilibrium reformer. Meanwhile direct oxidation has been achieved in the laboratory, as mentioned in Section A.3.2 (route 1).

Accordingly, the calculation for this simpler direct route, Figure A.5, is as follows:

$$CH_4 + 2O_2 \Leftrightarrow CO_2 + 2H_2O$$

See Figure A.5 also for half reactions, from which $\nu = 8$.
Using the JANAF tables for $T_o = 298.15K$, $P_o = 1$ bar:

CH_4 $S = 186.251, \Delta_f G = -50.768,$ $\log K_f = 8.894$

O_2 $S = 205.147, \Delta_f G = 0, \log K_f = 0$

$2O_2$ $S = 410.294, \Delta_f G = 0, \log K_f = 0$

CO_2 $S = 213.795, \Delta_f G = -394.389, \log K_f = 69.095$

H_2O $S = 65.215, \Delta_f G = -237.141, \log K_f = 41.546$ (real fluid, in JANAF tables, Chase et al. 1998)

$2H_2O$ $S = 130.430, \Delta_f G = -474.282, \log K_f = 83.092$

Hence

$$\Delta S = -186.251 - 410.294 + 213.795 + 130.430 = -252.32$$

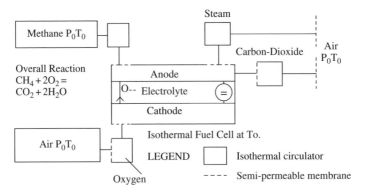

1] The Oxidation is Isothermal at To, leading to the calculation of the Methane Chemical Exergy
2] The practical reaction would be catalysed at about 600°C, in a suitable SOFC.
3] Cathode half reaction, $2O_2 + 8e- = 4O-$
4] Anode half reaction, $CH_4 + 4O-- = CO_2 + 2H_2O + 8e-$
5] Dissociation and Cell Equilibria, $H_2 + 1/2\ O_2 = H_2O$, $CO + 1/2\ O_2 = CO_2$

Figure A.5 Equilibrium calculation route for direct oxidation of methane

and

$$\Delta_r G = 50.768 - 394.389 - 474.282 = -817.903$$
$$\log_{10} K_r = -8.894 + 69.095 + 83.092 = 143.293$$

Hence

$$V_n = -\Delta G/vF = 1000 \times 817.903/8 \times 96484 = 1.059\,V$$

A.3.11 Remarks on Equilibrium of Direct Electrochemical Methane Oxidation

For hydrogen and carbon monoxide the cell equilibrium was identical to the single product, water or carbon dioxide, dissociation into the reactants.

For methane oxidation, at equilibrium, the two products within the fuel cell, are CO_2 at about 0.97 bar and H_2O (steam/water) at about. 0.03 bar, the latter being the saturation pressure corresponding to 298.15 K. There is a reversible reaction between the two products and the two reactants, methane and oxygen. The result of the right-handed equilibrium calculation is going to be sparse reactants, relative to the high concentration of products. The distinction is again being made here between non-equilibrium, low-power-yield combustion, where equilibrium constants are of no avail, and isothermal oxidation in a fuel cell, where equilibrium constants are relevant, and power yield is relatively high.

The fuel cell oxidation equilibrium equation is

$$0 = -CH_4 - 2O_2 + CO_2 + 2H_2O$$

from which a simplistic equilibrium constant, on the lines of Eqn (A.1), may be derived by the procedures of Atkins (1995), pp. 278–279. Thus

$$K = \{[F(CO_2)/P_\theta] \times [F(H_2O)/P_\theta]^2\}/$$
$$\{[F(CH_4)/P_\theta] \times [F(CH_4)/P_\theta] \times [F(O_2)/P_\theta]^2\} \quad \text{(A.5)}$$

A.3.12   Equilibrium Concentrations for Direct Methane Oxidation

Applying the principle of conservation of mass, with z as the small fraction of methane in kmol present in the fuel cell equilibrium mixture, the balanced oxidation equilibrium is

$$zCH_4 + 2zO_2 + (1-z)CO_2 + (2-z)H_2O$$

The total number of moles in the mixture, n, the amount of substance, is

$$n = z + 2z + 1 - z + 2 - z = z + 3$$

The reactants, z methane and $2z$ oxygen are in stoichiometric ratio. With small z, the carbon dioxide and water products predominate. Their joint decomposition matches their formation from the two reactants. Hence the molar composition of the mixture is

$$\gamma_{CH_4} = z/(z+3), \quad \gamma_{O_2} = 2z/(z+3), \quad \gamma_{CO_2} = (1-z)/(z+3),$$
$$\gamma_{H_2O} = (2-z)/(z+3)$$

For very small z, $\gamma_{CH_4} = z/3$, $\gamma_{O_2} = 2z/3$, $\gamma_{CO_2} \approx 1/3$, $\gamma_{H_2O} \approx 2/3$ (based on number of product gas molecules).

The realistic figures are $\gamma_{CO_2} = 0.97$ and $\gamma_{H_2O} = 0.03$, and the large difference between the two pairs of γ values comes from non-differentiation by the equilibrium constant calculation between the actual two-phase H_2O product and an ideal gas:

$$\log_{10} K_r = -143.293, \quad \ln\ K = -330.004 \text{ and } K = 5.105\ e^{-144}$$

Equation (A.5) then becomes at P_0/T_0 (when for very small z, $\gamma_{CH_4} = z/3$, $\gamma_{O_2} = 2z/3$)

$$K = (\gamma_{CO_2}) \times (\gamma_{H_2O})^2/(\gamma_{CH_4}) \times (\gamma_{O_2})^2 = (1/3 \times 4/9)(z/3 \times (2z/3)^2)$$
$$= (4/27)(4z^3/27) = z^{-3} = 5.105e^{-144}$$

Hence

$$z = (5.105\,e^{-144})^3 = (133.04\,e^{-48}) = 1.3304\,e^{-50}$$

and $\gamma_{CH_4} = 4.43\,e^{-49}$, $\gamma_{O_2} = 8.86\,e^{-49}$, $\gamma_{CO_2} = 1/3$ (0.03: two-phase basis) and $\gamma_{H_2O} = 2/3$ (0.97: two-phase basis).

A.3.13   Circulator Powers, $\Delta G = -RT\ln(P_i/P_f)$

For CH_4,

$$\Delta G = -[8.315 \times 298.15 \times \ln(1/(4.43\,e^{-49})]/1000$$
$$= -8.315 \times 298.15 \times 111.34/1000 = -276.025$$

For $2O_2$,

$$2\Delta G = -2[8.315 \times 298.15 \times \ln(0.21/8.86\,e^{-49})]/1000$$
$$= -2 \times 8.315 \times 298.15 \times 109.084/1000 = -540.864$$

For $2H_2O$, using Mollier chart extrapolation,

$$\Delta G = -2 \times 298.15(9.21 - 8.57) \times 18/1000 = -6.87$$

Note that the steam/water must be at saturation temperature 298.15 K and saturation pressure 0.03 bar. A mixture of perfect gases would have shared the pressure pro rata to their mole concentrations. Hence CO_2 pressure $= 0.97$ bar.

For CO_2,

$$\Delta G = -8.315 \times 298.15 \times \ln(0.97/0.0003)/1000$$
$$= -8.315 \times 298.15 \times 10.384/1000 = -25.74$$

Hence

$$\text{Total circulation power} = -[276.025 + 580.864 + 6.87 + 25.74]$$
$$= -889.5$$

and thus

Methane chemical exergy $= -889.5 - 817.9$

$$= -1707.4 \, kW \, s \, mol^{-1} \, \textbf{(direct oxidation)}$$

The reformer route gives $-990 - 817.9 = -1807.9$ above (Section A.3.9). The mismatch in circulator powers is acceptable for two disparate calculations involving approximate equilibrium constants approximate concentrations and logs of very large pressure ratios.

A.3.14 Conclusions

The conclusions are as follows:

1. The thermodynamic basis of the calculation of the maximum possible work potential or chemical exergy of reversible and irreversible chemical reactions is explained and discussed. Combustion is asserted to be fundamentally irreversible. It is a non-equilibrium uncontrollable chain reaction with hot branches, in a cool milieu, and a limited work output proportional to Carnot efficiency × calorific value (Barclay, 2002).

2. On the other hand, the charge transfer isothermal chemical reaction of the fuel cell is reversible at zero throughput. The reaction can be slowed down towards zero rate and equilibrium by controlling its throughput. The output power, including that of the circulators, maximises at $P_0 T_0$, in a special apparatus, which for hydrogen and carbon monoxide comprises three membrane-assisted circulators or concentration cells, necessary to provide reactants and evacuate product at equilibrium concentrations,

into and from the fuel cell. The equilibrium concentrations are the same as those of the dissociated product (steam/water or carbon dioxide), namely rarefied fuel and oxygen, with concentrated product.

3. Methane fuel has alternative and more complex equilibrium electrochemical oxidation routes and two of these are examined because of their resemblance to practical routes. The first is via a reformer and the second is via direct oxidation, now achievable in the laboratory. Both analyses involve approximate equilibrium constants, but the second direct oxidation calculation route is seen to be more in error, and the numerical answer less accurate, than that of the reformer route.

4. The $P_0 T_0$ fuel chemical exergies of hydrogen and of carbon dioxide are calculated as 477 and 534.5 $W\,s\,mol^{-1}$. Above $P_0 T_0$ the work potential falls with temperature and, in the case of hydrogen, with pressure. Because of the foregoing, the chemical exergy of a fuel is accurately calculable only at standard conditions, namely $P_0 = 1$ bar, $T_0 = 298.15$ K.

5. That part of the fuel chemical exergy due to circulator power (about half) cannot be realised, because of the very large pressure ratios between the inlet and outlet of the circulators, be they mechanical or concentration cells. Moreover, adequate permselective membrane technology is not available. Practical air-breathing systems preclude circulators, so that the present-day fuel cell has a limitation comparable in size with the fundamental Carnot limitation. However, the realisable fuel chemical exergy (work potential) is nevertheless much larger than that based on the Carnot limited enthalpy of combustion (energy). Fuel cells are still very much worthwhile, by virtue of their large potential power output, relative to combustion. The development potential available to combustion engineers is a function of the enthalpy of combustion $\times$ Carnot efficiency, with a limitation at high temperatures due to dissociation. The development potential available to fuel cell engineers is the fuel chemical exergy, a much larger quantity.

6. The equilibrium thermodynamics of a hydrogen fuel cell are further compared with those of a carbon monoxide fuel cell. At equilibrium both cells can be described as having a high vacuum of reactants and a high concentration of products.

The comparison highlights the difference between the non-ideal hydrogen/steam/water case and the ideal carbonmonoxide/carbondioxide case. The difference can be detected only if fugacity-based calculations as displayed in the introduction to this book are made using the JANAF tables, (Chase *et al.*, 1998). The equilibrium concentrations, the equilibrium constant and the Nernst potential difference V_n, in the hydrogen case, are a function of both pressure and temperature. V_n declines with pressure. In the carbon monoxide perfect gas case, the same variables are a function of temperature only. The pressure coefficient is zero.

7. The equilibrium of a hydrogen or carbon monoxide fuel cell operating at high temperature and pressure is defined using a flow sheet, which connects the cell to a fuel store at standard conditions, and to the environment, via combined isentropic and isothermal circulators and a Carnot cycle.

   The associated calculations show that maximum theoretical power derived from the fuel chemical exergy is obtainable only at standard conditions. The fuel work potential breaks down into fuel cell electrical power, circulator power and Carnot cycle power. The circulator power is large, resulting from huge pressure ratios associated with the very low (high-vacuum) reactant concentrations. The sum of the fuel cell electrical power and the Carnot power, namely the 'combined power', is roughly constant for all operating conditions, and that sum is the currently pursued practical objective, using air-breathing fuel cells without circulators.

8. The fuel chemical exergy (work) is shown to be much greater than the calorific value (energy), and indeed the large numerical difference between the two, via a simple calculation, represents the great theoretical advantage of the fuel cell.

9. Hydrogen and carbonmonoxide are fuels which must be made at thermodynamic and economic cost. A principal industrial route is via the fired steam reform of natural gas, a highly irreversible process. The related thermodynamically reversible route to methane reform, and electrochemical oxidation, Figure A.3, is examined. An electrically driven electrochemical reformer at standard conditions is the model. The reformer supplies a pair of fuel cells separately utilising carbon monoxide and hydrogen. The thermodynamic data confirm that there is plenty of electricity available

to drive the reformer and still provide a substantial electrical system output. The latter net output, the methane chemical exergy, represents the theoretical maximum output by Faradaic means from methane, $818 + 890 = 1708\,\mathrm{kW\,s\,mol}^{-1}$.

10. The argument can be laboriously extended to natural gas of variable analysis, but that task is of marginal importance, and is not within the scope of this appendix, and book.

11. The direct methane oxidation route, Figure A.5, is also examined. At $P_0 T_0$ the products are carbon dioxide and two-phase steam/water. The water is an addition to the water electrolyte. To maintain balance steam is extracted by a circulator and passed to the environment, but the simplistic equilibrium coefficient used in the calculation is based on perfect gases, and does not predict the two-phase steam/water The resulting methane chemical exergy figure is $818 + 990 = 1808\,\mathrm{kW\,s\,mol}^{-1}$. A similar but smaller error occurs in the reformer route, and hence the preferred value is $1708\,\mathrm{k\,W\,s\,mol}^{-1}$.

    The numbers 890 and 990 are relatively doubtful circulator powers, while 818 is electrochemical power.

12. By raising the operating pressure and temperature, the output of equilibrium fuel cells is shown to be reduced. The output includes electrical power from the fuel cell, power from the isentropic and isothermal circulators, and power from the Carnot cycle. No artificial credit is given for producing combined heat and power by the popular but erroneous addition of power and heat, which have different units, to produce mythical high efficiencies. The addition of similar units (e.g. power plus power, or heat plus heat) is the valid procedure, in line with Joule's irreversible experiment, from which $1\,\mathrm{W\,s} \gg 1\,\mathrm{J}$. The $\gg$ sign can never mean $=$. In particular the stirred liquid of the experiment cannot be unstirred.

13. The analyses here differ from those of Gardiner (1996), Kotas (1995) and Moran and Shapiro (1993) because of the use of the fugacity calculations from the JANAF tables (Chase *et al.*, 1998), and, more importantly, because the contents of the isothermal enclosure of the fuel cell are at concentrations determined by the equilibrium constant (high vacuum of reactants, high concentration of products). The introduction of a Faradaic reformer is new.

14. Finally, the author acknowledges his indebtedness to Gardiner (1996), whose paper he heard at its first presentation, and reread many times, before being certain that the concept of isothermal oxidation therein did not gel with his own, because of not using equilibrium concentrations.

A.4   ELABORATION OF FIGURES A.4 AND A.5, THE EQUILIBRIUM METHANE OXIDATION ROUTES

Figures A.4 and A.5 were drawn initially in the pursuit of a calculation route for the chemical exergy of methane. For that purpose the two equilibrium diagrams each provide a reversible route to power production from methane. Each process gave a confidence-raising, similar answer for the chemical exergy of methane.

In Figure A.4 water and methane, each proceeding from store, while being driven by circulators via perm-selective membranes, are reacted isothermally in a notional and conceptually quite new isothermal reformer resembling a fuel cell bounded by perm-selective membranes. The circulators are isothermal concentration cells with Nernst potential differences.

The water reaction is $2e^- + 2H_2O = H_2 + 2OH^-$.

The methane reaction is $2CH_4 + 2OH^- = 2CO + 5H_2 + 2e^-$.

The reformer outputs of carbon monoxide and hydrogen are reacted with atmospheric oxygen in separate fuel cells to produce power. At the cost of power from the hydrogen fuel cell, hydrogen may be removed for insertion into the fuel tanks of vehicle fuel cells. If the hydrogen removal is permanent the size of the hydrogen fuel cell can be reduced. The set up is then a **hydrogen mine**.

The propulsion of vehicles by fuel cell may well require the development of a hydrogen mine. Such a plant would serve the needs of organisations such as Air Products or Methanex, who might even be its developers! Combustion-heated reformers would be displaced from the hydrogen industry.

Present-day vehicle demonstration buses use irreversible and incomplete electrolyser hydrogen sources. The output of carbon dioxide from the stack of the power plant supplying the electrolyser will exceed the pollution saved at the bus exhaust, a non-viable situation.

A possible technology to be applied to the hydrogen cell in a practical version of Figure A.4, subject to semi-permeable membrane and concentration cell availability, would be that of the Ballard PEMFC. But where carbon monoxide is involved, in the reformer and carbon monoxide cell, other technology would be needed. The development problem is a big one, to be minimised by careful choice of initial design.

Because the IT/SOFC integrates well with the gas turbine so as to burn fuel not consumed by isothermal oxidation, and because the gas turbine compressor provides the means to elevate the system temperature to ion conduction level for start-up purposes, the IT/SOFC is a promising choice, with a wide selection of competing types.

For Figure A.5, which is in no sense a hydrogen mine but a pure power plant of very high productivity, the gas-turbine-integrated IT/SOFC is additionally attractive because it accommodates any of the special techniques for methane oxidation (Perry Murray *et al.*, 1999; Park *et al.*, 1999; Gorte *et al.*, 2000).

A.5   PRACTICAL POWER PRODUCTION FOR THE FUTURE

In Figure A.6, an indication is given of the nature of a practical plant for the isothermal oxidation of methane from natural gas. The heavy hydrocarbons are removed from the natural gas and sent to the combustion chamber of a gas turbine, to be burnt in air with the hot mixed exhaust from the fuel cell of unused fuel, steam and carbon dioxide.

The compressor outlet air temperature is matched to the IT/SOFC. Since the plant would be a large base load system it would have a multipressure steam plant to use its exhaust heat.

The plant in Figure A.4 can be dealt with in exactly the same way. The reformer and the two fuel cells would be elevated to IT/SOFC conditions, as in Figure A.6. All surplus fuel, heavy hydrocarbons and unoxidised fuel from the three plant sections, together with three hot exhausts, would be swallowed by a gas turbine combustion chamber as above. That would yield a controllable plant, subject to availability of semi-permeable membranes and of isothermal concentration cells, appropriate to IT/SOFC temperatures and gas turbine pressure.

Laboratory semi-permeable membranes with a solid oxide basis have appeared in Dyer *et al.* (2000) bringing forward the achievement of practicality as aimed at in Figure A.6.

The main plant output from fuel chemical exergy would be a function of the amount of fuel isothermally oxidised in the fuel cell, with a minority contribution from the calorific value of the fuel burnt in the

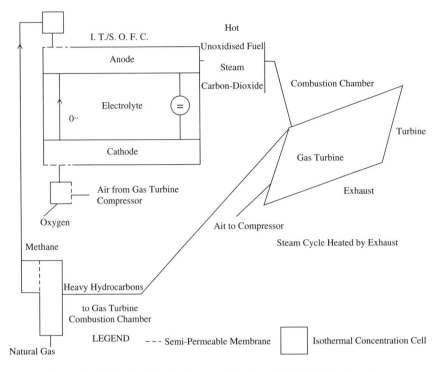

Figure A.6 Practicality in direct oxidation via IT/SOFC technology

gas turbine. The gas turbine contributes valuable controllability via its compressor which, on mains power, is able to raise the temperature of the cell to ion conduction level for start-up. The designer would seek to maximise isothermal oxidation.

A.5.1 Concentration Cells for Future IT/SOFC/Gas Turbine Power

The flow sheet in Figure A.6 then requires the addition of isothermal concentration cells to give their contribution to the power, and increase the Nernst potential difference. These would be immediately adjacent to the fuel cell in the methane and oxygen supply lines. The rearranged flow sheet is shown as Figure A.7.

The fuel cell and the two concentration cells each generate Nernst potential differences, to be put in a series external circuit so as to enhance current and power production. The natural gas is separated into methane and heavy hydrocarbons, the latter to be burned by the gas turbine. The gas turbine is shown as a temperature–entropy diagram as in Figure A.6.

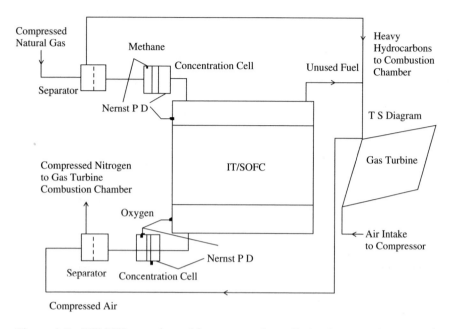

Figure A.7 IT/SOFC complete with concentration cell circulators and integrated with gas turbine

Also burned in the gas turbine combustion chamber is the excess fuel and oxygen which the isothermal oxidation process of the fuel cell cannot consume. The gas turbine compressor takes in atmospheric air, which is separated into oxygen and nitrogen: oxygen for the fuel cell and compressed nitrogen to contribute its exergy to the gas turbine combustion chamber, in which it is heated for expansion in the turbine.

In the absence of developed components, the performance of the flow sheet cannot be calculated. It shows the problems which need to be tackled before the fuel cell industry can exploit its full potential.

A.5.2 Direct Low-Temperature Coal Fuel Cell

Harvard University announced in *IEE Power Engineer* of October/November 2005 a new fuel cell concept, which articulately uses charge exchange processes and therefore isothermal oxidation. That is a major leap forward for the industry. See also the two documents by Weibel *et al.* (2005a, b) for a much more detailed description.

It is a surprise that Figure A.8 shows the cell bridged by its internal resistance, so that much information is missing. The data would be required over the range of Figure 6.5, from open-circuit to the

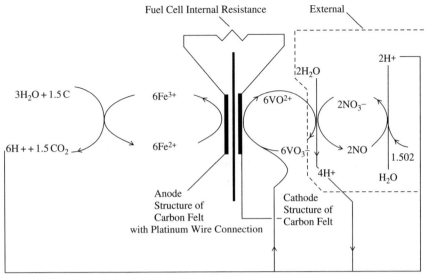

Electricity from Coal. Harvard University Experiment

Figure A.8   Harvard coal fuel cell

voltage–current droop region. The disposal of carbon dioxide and steam is not shown.

The authors react Fe^{3+} with a low-grade, acid, sub-bituminous coal (SBC) slurry to produce carbon dioxide and Fe^{2+}. The latter ion can then be harnessed as the anode of the cell, with the oxidation reaction Fe^{2+} to Fe^{3+} coupled to the reduction of vanadium dioxide at the cathode of the fuel cell. The first primitive device is said to be 7% efficient, but that is based on the fuel calorific value, asserted to be irrational in this book and in (Barclay, 2002). An alternative catalyst is mooted to start the path to improvement. The fate of the used slurry is not discussed. It could be burnt or buried, depending on economics.

In Weibel *et al.*, (2005a, b) the $100\,^{\circ}C$, fraction of 1 litre, experimental cell is described in detail, and its results are given. The cell shares features and terminology with Regenesys, in that it has an anolyte (5 M sulphuric acid, saturated with iron suphate and from 0.3 to 4g of SBC, separated by a Nafion 112 membrane, from the catholyte, VO_2^+ and VO^{2+} (total concentration 1 M) in 5 M sulphuric acid. The anode and cathode electrode structures are both of carbon felt. The cell has incompressible reactants, but its products, CO_2 and H_2O, at 100°C are both compressible. In an equilibrium diagram of the cell there would have to be, to avoid irreversible diffusion, an isothermal concentration cell to expand the products to their atmospheric concentrations, thereby generating additional Nernst potential difference. The cell development, however, has been empirical, and comprises carefully assembled experimental data for a system which, in the customary world fuel cell industry way, does not have an underlying basic thermodynamic theory, as provided by this book.

The route to developing a new fuel cell of such great potential importance must, this book asserts, entail drawing an equilibrium diagram, followed by decisions about those features of the equilibrium system which are unacceptable, or unachievable, succeeded by acceptance of the resulting calculated irreversibilities, and reduced performance related to the fuel chemical exergy, which can be guessed as twice that of CO, that is about $1070 \text{kW s mol}^{-1}$.

The reader can but keep in touch with the developments in the industry, using the means which this book points out, notably the US Patent Office and scholar.google.

Appendix B

Patent Search Examples

Example 1
Call up www service from your Internet service provider. Open
http://www.uspto.gov/patft/index.html/
Go to Pat Num. Enter 5527363. Text of patent comes up.
Select Images. Select Go to Page 4. Flow Plate Illustration 4A comes up.

Example 2
Go to Advanced Search, Patents Granted.
Enter AN/"acumentrics"
All years,
Eight Hits come up: Select No 1; Read text; or select images: there is an
image on the front page of the text.

Example 3
The same procedure with Ballard Power Systems instead of Acumentrics
produces 91 hits. Some easement of the task could come from searching
separate years. The task is reduced because the firm is specialised. The
company may have changed its name, resulting in earlier patents being
left out.

These are easy examples, because, for Example 1, the patent number is
provided; and for Example 2, there are only eight hits, because the firm
is new and specialised.

Fuel Cells, Engines and Hydrogen – An Exergy Approach   Frederick J. Barclay
© 2006 John Wiley & Sons, Ltd

Courses on patent searching are provided at major libraries, for example the British Library at St Pancras, London. At that library there is a vast paper and electronic system to be probed:

- UK Patent Office: http://www.patent.gov.uk.

- European Patent Office: http://www.ep.espacenet.com.

- VRB Power Systems: http://www.vrbpower.com/ for information on Regenesys.

- Google: http://www.scholar.google.com/ is a major new resource for fuel cell references. The search words 'fuel cell' throw up 41 000 references which can be sorted in various ways. References are updated continuously so that the unavoidable task of keeping up to date is lightened.

- ScienceDirect: http://www.sciencedirect.com/ is a recently organised path giving access to scientific journals.

- See http://www.fuelcelltoday.com/ for lists of articles and new patents on fuel cells.

Appendix C

List of Web Sites

http://www.abb.com   Asea Brown Boveri Group, suppliers of inverter equipment, for fuel cells.

http://www.acumentrics.com   Acumentrics Inc., USA.

http://www.adelan.co.uk   Adelan Ltd UK.

http://www.airproducts.com   Air Products Inc.

http://www.ballard.com   Ballard Power, Canada.

http://www.cerespower.com   Ceres Power Ltd UK.

http://www.cfcl.com.au   Ceramic Fuel Cells Ltd, Australia.

http://www.cfcl.com.au/Links/CFCL_40AlbertRd_Deal_%20050715.pdf/   CFCL/UK European Office, Regus House, Heron's Way, Chester Business Park, Chester, CH4 9QR.

http://www.cropper.com/tech_fibre_products.html   Components of MEAs made for Johnson Matthey Fuel Cells, see below at http://www.jmfuelcells.com.

http://www.delphion.com   Patent information.

http://www.denora.com   Denora Fuel Cells, Italy.

http://www.dti.gov.uk/NewReview/   Department of Trade and Industry, UK.

http://www.dti.gov.uk/renewable/index.html   Department of Trade and Industry, UK.

http://www.duvera.com   Duvera Fuel Cells, USA. Partners with Renault for Scenic Car Project.

http://www.ecn.nl   (includes InDEC, Innovative Dutch Electro-Ceramics Company).

http://www.exergysource.com   F J Barclay's web page.

http://www.fuelcellenergy.com   Fuel Cell Energy Inc., USA.

http://www.fz-juelich.de/iwv/e-iwv3.htm   Forschungs Zentrum Julich-Fuel Cells.

http://www.gasification.org/resource/library/libray.html   Library of UK. Gasification Organisation.

http://www.globalte.com   or   globalte.com/fcprogramme-frames.htm   Global Thermoelectric, Canada.

http://www.graftech.com   Graftech or Advanced Energy Products (Flexible Graphite), USA.

http://www.hexis.com   Sulzer Hexis Fuel Cells.

http://www.icheme.org/learning   Institution of Chemical Engineers, UK.

http://www.jmfuelcells.com   Johnson Matthey Fuel Cells, UK. Makes all MEAs for Ballard Power.

http://www.methanex.com   Methanex Co., Canada.

http://www.mmc.co.jp/English/top_e.html/   Mitsubishi Materials Corp., Japan.

http://www.mtu-online.com   Successors to MAN Friedrichshafen.

http://www.murata.co.jp   Murata Manufacturing Co., Japan.

http://www.sti.nasa.gov/casitrs.html   NASA Technical Reports Server.

http://www.nature.com   Nature Magazine, London.

http://www.newspress.co.uk   for Honda news.

http://www.ngk.co.jp   NGK Insulators, Japan.

http://www.nuvera.com   Nuvera Fuel Cells, Italy.

http://www.pepublishing.com   Professional Engineering Publishing Ltd.

http://www.pg.siemens.com   Siemens Power Generation, Germany.

www3.interscience.wiley.com/cgi-bin/home/26377/renault.com   Renault Cars. Partners with Du Vera in Renault Scenic Fuel Cell Car Project.

http://www.pwc.com   Price Waterhouse Coopers Accountants, source of Annual Fuel Cell Industry Survey.

http://www.rolls-royce.com   Rolls-Royce Fuel Cells Ltd, UK.

http://www.netl.doe.gov/seca   National Energy Technology Laboratory, USA.

http://www.thecarbontrust.co.uk   The Carbon Trust is interested in CO_2 reduction and climate change mitigation.

http://www.tnwtudelft.nl   Technical University of Delft, the Netherlands.

http://www.tokyo-gas.co.jp   Tokyo Gas Company, Japan.

http://www.uspto.gov/patft/index.html   US Patent Office.

http://www.utcfuelcells.com   UTC Fuel Cells, USA.

http://www.victrex.com   Victrex Company Ltd UK (allied with Ballard
   Power).
http://www.wiley.com   John Wiley & Sons.
http://www.wiley-VCH.de   Wiley Germany.
http://www.xantrex.com   Xantrex Technology, suppliers of inverter
   equipment to the market.

Bibliography

Adams A M *et al.*, 1963, *The High Pressure Hydrogen Oxygen Cell*, Chemical Technology Vol. 1, Academic Press, New York.

Agnew G D, 2001, The Rolls Royce development programme for pressurised hybrid fuel cell systems. In *Seventh Grove Fuel Cell Symposium, London*, Article in Delegates Manual.

Akikusa J *et al.*, 2001, Development of a low temperature operation solid oxide fuel cell. *Journal of the Electrochemical Society*, **148**, A1275–A1278.

Appleby A J, 1994, Fuel Cell Electrolytes: Evolution, Properties, and Future Prospects. *Journal of Power Sources*, **49**, 15–34.

Atkins P W, 1995, *Physical Chemistry*, 5th Edition, Oxford University Press, Oxford.

Bacon F T, 1969, Fuel Cells, Past, Present, and Future *Electrochimica Acta*, **14**, 569–585.

Badwal S, 1996, Fuel Cell Interconnector, WO 96/28855 A1.

Badwal S *et al.*, 1997, An Electrical Interconnector for a Planar Fuel Cell, WO 9735349 A1.

Ballhausen, A, 2001, A Demonstration of European Micro-CHP Fuel Cell Systems, In *Seventh Grove Fuel Cell Symposium, London*, Poster.

Bance P *et al.*, 2004, Spinning out a fuel cell company from a UK university – 2 years of progress at Ceres Power. *Journal of Power Sources*, **131**(1–2), 86–90.

Baozhen Li *et al.*, 2001, Solid Oxide Fuel Cell Operable over Wide Temperature Range. US Patent 6,207,311.

Barclay F J, 1995, *Combined Power and Process – An Exergy Approach*, Professional Engineering Publishing, London.

Barclay F J, 1998, *Combined Power and Process – An Exergy Approach*, 2nd Edition, Professional Engineering Publishing, London.

Barclay F J, 2002, Fundamental thermodynamics of fuel cell, engine, and combined heat and power system efficiencies. *Proceedings of the Institution of Mechanical Engineers, Part A, Journal of Power and Energy*, **216**, 407–417.

Bard A J and Falkener L R, 1980, *Electrochemical Methods, Fundamentals and Applications*, John Wiley & Sons, Inc., New York.

Bartolozzi M, 1989, Development of Redox Flow Batteries – A Historical Bibliography. *Journal of Power Sources*, **27**, 219–234.

Basel R A and Pierre J F, 1995, Pressurised Fuel Cell Testing. In *Proceedings of the Fuel Cells Review Meeting*, Ed. T J George, US Department of Energy, Morgantown, WV.

Batawi E, 1996, High Temperature Fuel Cell. US Patent 5,691,075.

Batawi E, 1999, Battery with High Temperature Fuel Cells. US Patent 5,902,692.

Batawi E, 2001, Electrochemically Active Element for a High Temperature Fuel Cell. US Patent 6,232,009.

Batawi E *et al.*, 1999, High Temperature Fuel Cell with a Thin Film Electrolyte. US Patent 5,923,368.

Batawi E *et al.*, 2001, Perowskite for a Coating of Interconnectors. US Patent 6,228,522.

Bockris J O'M and Srinivasan S, 1969, *Fuel Cells*, McGraw-Hill, New York.

Boehm G *et al.*, 2001, Method and Apparatus for Operating a Fuel Cell. WO 01/43216.

Borglum B P, 1999, Fuel cell Tubes and Method of Making Same. US Patent 5,993,985.

Brown K *et al.*, 2000, Gas Diffusion Substrates. WO 00/55933.

Campbell S A, 1999, Porous Electrode Substrate for an Electrochemical Fuel Cell. US Patent 5,863,673.

Campbell N A *et al.*, 2006, *Biology: Concepts and Connections*, 5th Edition, Pearson Benjamin Cummings, Upper Saddle River, NJ.

Carrette L *et al.*, 2000, Fuel Cells: Principles, Types, Fuels, and Applications. *Chem Phys Chem*, **1**, 162–193.

Chase M W *et al.*, 1985, *JANAF Thermochemical Tables*, 3rd Edition, American Chemical Society and American Institute of Physics, for the National Bureau of Standards, Thermal Group, Dow Chemical, Midland, MI.

Chase M W *et al.*, 1998, *JANAF Thermochemical Tables*, 4th Edition, American Chemical Society and American Institute of Physics, for the National Bureau of Standards, Thermal Group, Dow Chemical, Midland, MI.

Colbow K *et al.*, 2001, Method and Apparatus for Operating an Electrochemical Fuel Cell with Periodic Reactant Starvation. WO 01/01508.

Cooley G E and Nix K J, 1998, Method for the Fabrication of an Electrochemical Cell, Having Long Term Chemical Stability. US Patent 5,785,912.

Davidson B *et al.*, 2000, Large Scale Storage Solution: Regenesys™ Regenerative Fuel Cell. *Renewable Energy World*, Jan/Feb.

Domeracki W F *et al.*, 1995, Integrated Gas Turbine Solid Oxide Fuel Cell System. US Patent 5,413,879.

Donelson R *et al.*, 1998, A Fuel Cell Assembly. WO 98/57384 A1.

Doyon J *et al.*, 2003, The DirectFuel Cell™ stack engineering. *Journal of Power Sources*, **118**, 8–13.

Draper R *et al.*, 1987, High Thermal Conductivity Gas Feeder System. US Patent 4,664,986.

Duncan G C, Stephen H J S and Male E, 2000, Method for the Fabrication of Electrochemical Cells. US Patent 6,086,643.

Dyer P N *et al.*, 2000, Ion Transport Membrane Technology for Oxygen Separation and Syngas Production. *Solid State Ionics*, **134**, 21–33.

Eaves S *et al.*, 2004, A Cost Comparison of Fuel Cell and Battery Electric Vehicles. *Journal of Power Sources*, **130**(1–2), 208–212.

Foger K *et al.*, 2000, Fuel Cell System. WO 01/13452 A1.

Fletcher N J *et al.*, 1995, Electrochemical Fuel Cell Employing Air as the Oxidant and Coolant. US Patent 5,470,671.

Foster S, 1998, Routes to Interfacial Deposition of Platinum Microparticles for Solid Polymer Fuel Cells, PhD Thesis, University of Loughborough.

Frost J C *et al.*, 1999, Manufacture of Electrodes. US Patent 5,871,860.

Gardiner F J, 1996, Thermodynamic Processes in Solid Oxide and Other Fuel Cells. *Symposium S 455, Paper 3, Fuel Cells for Power and Propulsion*, IMechE, London.

Gardiner F J *et al.*, 1996, Solid Oxide Fuel Cell Stack. US Patent 5,486,428.

Gardiner F J *et al.*, 1997, Solid Oxide Fuel Cell Stack. US Patent 5,595,833.

George R A *et al.*, 2001, Single Module Pressurised Fuel Cell Turbine Generator System. US Patent (pending), WO 01/06589 A1.

Gibb P, 2001, Advances in Ballard Fuel Cell Technology - The Mk 900. *Seventh Grove Fuel Cell Symposium, London.*

Gibbs J Willard, 1961, *The Scientific Papers of J W Gibbs, Vol. 1, Thermodynamics*, Dover, New York.

Gibson R *et al.*, 1995, Solid Oxide Fuel Cell Stacking Arrangement. US Patent 5,460,897.

Gopinath R *et al.*, 2004, Development of a Low Cost Fuel Cell Inverter System with DSP Control. *IEEE Transactions on Power Electronics*, **19**(5), September, 1.

Gorte R J *et al.*, 2000, Anodes for Direct Oxidation of Dry Hydrocarbons in a Solid Oxide Fuel Cell. Advanced Materials, doi.wiley.com.

Grahame D C, 1957, I - Capacity of the Electrical Double Layer between Mercury and Aqueous Sodium Fluoride. II – Effect of Temperature and Concentration. *Journal of the American Chemical Society*, **LXXIX**, 2093–2098.

Grimble R E, 1988, Reforming of Fuel inside Fuel Cell Generator. US Patent 4,729,931.

Guelcher, S A *et al.*, 1998, Thermo-capillary Phenomena and Bubble Coalescence during Electrolytic Gas Evolution. *Journal of the Electrochemical Society*, **145**(6), 1848–1855.

Hoogers G *et al.*, 2002, *Fuel Cell Technology Handbook*, CRC Press, Boca Raton, FL.

Hotchkiss R, 2003, Coal Gasification Technologies. *Journal of Power and Energy*, 217(A1), 27–33.

Isenberg A O, 1983, Fuel Cell Generator. US Patent 4,395,468.

Jaffrey D, 1999, A Heat Resistant Steel. WO 99/25890 A1.

Johnson M C, 1998, Electrochemical Fuel Cell with an Electrode Structure Having an In-Plane Non Uniform Structure for Control of Reactant and Product Transport. US Patent 5,840,438.

Kendall K *et al.*, 2002, Fuel Cell Power Generating System. US Patent 6,358,640.

Knights S *et al.*, 2000, Method and Apparatus for Detecting a Leak Within a Fuel Cell. WO 00/39870.

Knights S *et al.*, 2001a, Fuel Cell Anode Structure for Voltage Reversal Tolerance. WO 01/15247.

Knights S *et al.*, 2001b, Method for Operating Fuel Cells on Impure Fuels. US Patent 6,210,820.

Koryta J, 1991, *Ions, Electrodes and Membranes*, 2nd Edition, John Wiley & Sons, Ltd, Chichester.

Koryta J, 1993, *Principles of Electrochemistry*, John Wiley & Sons, Ltd, Chichester.

Kotas T J, 1995, *The Exergy Method of Thermal Plant Analysis*, Krieger Malabar, FL.

Kumar G S *et al.*, 1995, High performance electrodes with very low platinum loading for polymer electrolyte fuel cells. *Electrochimica Acta*, 40, 285–290.

Kuo L *et al.*, 1999, Air Electrode Composition for Solid Oxide Fuel Cell. US Patent 5,932,146.

Lakeman J B *et al.*, 1999, The Regenerative Fuel Cell for Air Independent Power. In *UK Conference, Warship '99, Naval Submarines 6*.

Lamont G J *et al.*, 1998, Method and Apparatus for Detecting and Locating Perforations in Membranes Employed in Electrochemical Fuel Cells. US Patent 5,763,765.

Larminie J and Dicks A, 2000, *Fuel Cell Systems Explained*, John Wiley & Sons, Ltd, Chichester.

Lessner P, 1986, Kinetics of Aqueous Polysulfide Solutions, I and II. *Journal of the Electrochemical Society*, 133(12), 2510–2516 and 2517–2522.

Lessner P, 1988, Kinetics of Aqueous Polysulphide Solutions. *Journal of the Electrochemical Society*, 135(1), 258–259.

Liebhafsky H A and Cairns E J, 1968, *Fuel Cells and Fuel Batteries*, John Wiley & Sons, Inc., New York.

Linden D and Reddy T, 2002, *Handbook of Batteries*, 3rd Edition, McGraw-Hill, New York.

Lovell K and Page N, 1997, Membrane Electrolyte Technology for Solid Polymer Fuel Cells. In *ETSU F/02/00110/REP, Cranfield University*.

Marcus R A, 1964, Chemical and Electrochemical Electron Transfer Theory. *Annual Review of Physical Chemistry*, 15, 155–196.

Marcus R A, 1982, Electron Proton and Related Transfers. *Faraday Discussions of the Chemical Society*, 74, 7–15.

Masahiro Watanabe *et al.*, 1998, Polymer Electrolyte Membranes Incorporated with Nano-meter Size Particles of Pt and/or Metal Oxides: Experimental Analysis of the Self-Humidification and Suppression of Gas Crossover in Fuel Cells. *Journal of Physical Chemistry B*, **102**, 3129–3137.

McIntosh S and Gorte, 2003, Direct Hydrocarbon Solid Oxide Fuel Cells. Available from http://www.upenn.edu.

Mercuri R A, 2001a, Fluid Permeable Flexible Graphite Article with Enhanced Electrical and Thermal Conductivity. WO 01/54213.

Mercuri R A, 2001b, Graphite Article Useful as an Electrode for an Electrochemical Fuel Cell. WO 01/80343.

Mercuri R A, 2002, Fuel Cell Electrode Assembly with Selective Catalyst Loading. WO 02/27826 A1.

Mercuri R A *et al.*, 2002b, Fluid Permeable Graphite Fuel Cell Electrode with Enhanced Electrical and Thermal Conductivity, US Patent 6,486,686.

Minh N Q *et al.*, 1993, Monolithic Solid Oxide Fuel Cells with Integral Manifolds. US Patent 5,526,499.

Minh N Q and Takahshi T, 1995, *Science and Technology of Ceramic Fuel Cells,* Elsevier, Amsterdam.

Möbius H and Roland B, 1968, *Method of Producing Fuel Cells with Solid Electrolytes and Ceramic Oxide Electrode Layers.* US Patent 3,377,203.

Moran M J and Shapiro H N, 1993, *Fundamentals of Engineering Thermodynamics*, 2nd Edition, John Wiley & Sons, Ltd, Chichester.

Morita H *et al.*, 2001, The Influence of Operating Temperature on the Efficiency of Combined Fuel Cell and Power Cycle. *Journal of the Electrochemical Society*, **148**, A1051.

Morita H *et al.*, 2004, Types of fuel cell and factor analysis of performance. *Proceedings of the 2nd International Fuel Cell Conference*, Rochester, NY, p. 51.

Morrisey P J, 2001, Method of Operating a Fuel Cell. UK Patent Application GB 236,275.

Mortimer C E, 1975, *Chemistry, a Conceptual Approach*, Van Nostrand, New York.

Murray R C, 1983, Thermodynamics of Aqueous Sulphur Species. *Journal of the Electrochemical Society*, **130**(7), 866–869.

Neng You Jia, 2001, Method for Activating a Solid Polymer Electrolyte Fuel Cell. UK Patent Application, GB 2,362,500.

Park S *et al.*, 1999, Direct Oxidation of Hydrocarbons in a Solid Oxide Fuel Cell: 1 Methane Oxidation. *Journal of the Electrochemical Society*, **146**, 3603.

Park S *et al.*, 2000, Direct Oxidation of Hydrocarbons in a Solid Oxide Fuel Cell. *Nature*, **404**, 265–267.

Park S *et al.*, 2001, Tape Cast Solid Oxide Fuel Cells for the Direct Oxidation of Hydrocarbons. *Journal of the Electrochemical Society*, **148**, A443.

Perry Murray E *et al.*, 1999, A Direct Methane Fuel Cell with a Ceria Based Anode. *Nature*, **400**(12), 649–651.

Porterfield W W, 1993, *Inorganic Chemistry - A unified Approach*, 2nd Edition, Academic Press, San Diego.

Power Conversion Section, 2003, ABB Power Conditioning Systems, 80kW PEM Fuel Cell System, Hydrogenics Mississauga, Canada. Available from abb.com downloads.

Power Conversion Section, 2005a, Gate Drive for IGBT (Insulated Gate Bipolar Transistor). Available from abb.com downloads.

Power Conversion Section, 2005b, PCS 500 Power Converter Systems, with Power Electronics Building Blocks (PEBB's). Available from abb.com downloads.

Power Conversion Section, 2004, ABB Review 2, Electricity from the Store. Available from abb.com downloads.

Prater K B, 1990, The Renaissance of the Solid Polymer Fuel Cell. *Journal of Power Sources*, **29**, 239–250.

Price Waterhouse Coopers, 2005, Fuel Cell Industry Survey. Download available from pwc.com.

Ralph T R and Hogarth M P, 2002, Catalysis for Low Temperature Fuel Cells, Part 1, The Cathode Challenges. *Platinum Metals Review*, **46**(1).

Ralph T R *et al.*, 1997, Membrane Electrode Assemblies, MEA's, for Ballard Fuel Cells. *Platinum Metals Review*, **41**(3).

Ralph T R *et al.*, 2002, Catalysis for Low Temperature Fuel Cells, Part 2, The Anode Challenges. *Platinum Metals Review*, **46**(1).

Ralph T R *et al.*, 2003, Catalysis for Low Temperature Fuel Cells, Part 3, Challenges for the Direct Methanol Fuel Cell. *Platinum Metals Review*, **46**(1).

Remick R J and Ang P G P, 1984, Electrically Rechargeable, Anionically Active, Reduction/Oxidation, Electrical Storage System. US Patent 4,485,154.

Roberts J *et al.*, 2001, Method and Apparatus for Increasing the Temperature of a Fuel Cell with Polymer Electrolyte. WO 01/03215.

Rowland M, 1992, *Biology*, Thomas Nelson, Walton on Thames.

St Pierre J *et al.*, 2001, Integrated Fuel Cell and Pressure Swing Adsorption System. WO 01/47050.

St Pierre J *et al.*, 2001, Method and Apparatus for Increasing the Temperature of a Fuel Cell Stack. WO 01/48846.

Savadogo O and Essalik A, 1996, Effect of Platinum Particle Size on the Oxygen Reduction Reaction on $2\%Pt-1\%H_2WO_4$ in Phosphoric Acid . *Journal of the Electrochemical Society*, **143**(6), 1814–1821.

Schmickler W, 1993, Structure of Electrified Interfaces. In *Models for the Interface Between a Metal and an Electrolyte Solution*, Ed. J Lipkowski and P N Ross, VCH, Weinheim.

Schuler A, 2001, Plant with High Temperature Fuel Cells. US Patent 6,274,260.

Singh S, 2004, *Big Bang*, Clays, St Ives.

Singhal S C, 2000, Advances in Solid Oxide Fuel Cell Technology. *Solid State Ionics*, **135**, 305–313.

Somers E V and Isenberg E O, 1983, Fuel Cell Generator, and Method of Operating Same. US Patent 4,374,184.

Steele B C H *et al.*, 2000a, Ceramic ion conductors for fuel cells, oxygen separation and syngas production. *Ceramics: Getting into the 2000's*, Pt D, CIMTEC.

Steele B C H *et al.*, 2000b, Improving Gd doped ceria electrolytes for low temperature SOFC's. *Material Research Society Symposium Proceedings*.

Steele B C H *et al.*, 2001, The enabling technology for the commercialisation of fuel cell systems. *Journal of Materials Science and Engineering*, 39, 1112–1131.

Taylor J *et al.*, 2001, Cell Anode Structures for Voltage Reversal Tolerance. WO 01/15249.

Tillmetz W *et al.*, 2000, Fuel System with Improved Starting Capability. WO 00/30200.

Tolbert L M *et al.*, 2002, Charge Balance Control Schemes for Cascade Multilevel Converter in Hybrid Electric Vehicles. *IEEE Transactions on Industrial Electronics*, 49(5), 1058–1064.

Tseung A C, 1997, Fuel Cells: Past, Present and Future. University of Greenwich, Inaugural Lecture Series.

Vielstich W *et al.*, Eds, 2003, *Handbook of Fuel Cells*, 4 volumes, John Wiley & Sons, Ltd, Chichester.

Voss H H *et al.*, 1993, Method and Apparatus for Removing Water from Electrochemical Cells. US Patent 5,260,143.

Voss H H *et al.*, 1995, Anode Water Removal: A Water Management and Diagnostic Technique for Solid Polymer Fuel Cells. *Electrochimica Acta*, 40(3), 321–328.

Washington K B *et al.*, 1994, Laminated Fluid Flow Assembly for Electrochemical Fuel Cell. US Patent 5,300,370.

Weibel D B *et al.*, 2005a, Oxidation of Coal by Ferric Ion at 100°C as the Basis for a Coal Fuel Cell. *Angewandte Chemie*, 117, 5828–5832.

Weibel D B *et al.*, 2005b, Modeling the Anodic Half-Cell of a Low-Temperature Coal Fuel Cell. *Angewandte Chemie International Edition*, 44, 5682–5686.

Wilkinson D P *et al.*, 1984, Electrochemical Fuel Cell Assembly with Integral Selective Oxidiser. US Patent 4,482,680.

Wilkinson D P *et al.*, 1993, Light Fuel Cell Membrane Electrode Assembly with Integral Reactant Flow Passages. US Patent 5,252,410.

Wilkinson D P *et al.*, 1994, Solid Polymer Fuel Cells Incorporating Water Removal at the Anode. US Patent 5,366,818.

Wilkinson D P *et al.*, 1995, Method and Apparatus for Oxidising Carbon Monoxide in the Reactant Stream of an Electrochemical Fuel Cell. US Patent 5,432,021.

Wilkinson D P *et al.*, 1996a, Embossed Fluid Flow Field Plate for Electrochemical Fuel Cells. US Patent 5,521,018.

Wilkinson D P *et al.*, 1996b, Method of Fabricating an Embossed Fluid Flow Field Plate. US Patent 5,527,363.

Wilkinson D P *et al.*, 1997, Method and Apparatus for Reducing Crossover in an Electrochemical Fuel Cell. US Patent 5,672,439.

Wilkinson D P *et al.*, 1998, Electrochemical Fuel Cell Stack with Concurrent Flow of Coolant and Oxidant Streams, and Countercurrent Flow of Fuel and Oxidant Streams. US Patent 5,773,160.

Wilkinson D P *et al.*, 1998, Electrode. US Patent 5,795,669.

Wilkinson D P *et al.*, 1999, Method and Apparatus for Reducing Reactant Crossover in a Liquid Feed Electrochemical Fuel Cell. US Patent 5,874,182.

Wilkinson D P *et al.*, 1999, Electrochemical Cell with Fluid Distribution Layer Having Integral Sealing Capability. US Patent 5,976,726.

Wilkinson D P *et al.*, 2000a, Electrochemical Fuel Cell Membrane Electrode Assembly with Porous Electrode Substrate. US Patent 6,060,190.

Wilkinson D P *et al.*, 2000b, Impregnation of Microporous Electrocatalyst Particles for Improving Performance in an Electrochemical Fuel Cell. US Patent 6,074,773.

Wilkinson D P *et al.*, 2000c, Method and Apparatus for Controlling the Temperature within an Electrochemical Fuel Cell. WO 00/33407.

Wilkinson D P *et al.*, 2001, Fuel Cell Separator Plate with Discrete Fluid Distribution Features. WO 01/48843.

Williams M, 2002, *Fuel Cell Seminar, 18–21 November, Palm Springs, CA*, Abstracts.

Wolfe D *et al.*, 1999, Hybrid Electrical Power System. US Patent 5,968,680.

Xuezhong Du *et al.*, 2001, Performances of proton exchange membrane fuel cells with alternate membranes. *Physical Chemistry Chemical Physics*, **3**, 3175–3179.

Zhang J *et al.*, 2001, Ionomer Impregnation of Electrode Substrates for Improved Fuel Cell Performance. US Patent 6,187,467.

Zhang J *et al.*, 2001, Fuel Cell (Methanol) Sensor with Small Load Resistance and High Oxidant Supply. WO 01/35478.

Zito R, 1996, Electrochemical Apparatus for Energy Storage and/or Power Delivery Comprising Multi-Compartment Cells. US Patent 5,496,659.

Zito R, 1997, Process for Energy Storage and/or Power Delivery with Means for Restoring Electrical Balance. US Patent 5,612,148.

Zito R and Kunz L J, 1972, Method of Operating a Fuel Cell Using Sulphide Fuel. US Patent 3,920,474.

Zymboly G E, 1988, High Bulk Self Supporting Electrode with Integral Gas Feed Conduit for Solid Oxide Fuel Systems. US Patent 4,751,152.

Index